AF581646

Advanced Approaches, Business Models, and Novel Techniques for Management and Control of Smart Grids

Advanced Approaches, Business Models, and Novel Techniques for Management and Control of Smart Grids

Editors

Pierluigi Siano
Miadreza Shafie-khah

MDPI • Basel • Beijing • Wuhan • Barcelona • Belgrade • Manchester • Tokyo • Cluj • Tianjin

Editors
Pierluigi Siano
University of Salerno
Italy

Miadreza Shafie-khah
University of Vaasa
Finland

Editorial Office
MDPI
St. Alban-Anlage 66
4052 Basel, Switzerland

This is a reprint of articles from the Special Issue published online in the open access journal *Energies* (ISSN 1996-1073) (available at: https://www.mdpi.com/journal/energies/special_issues/Management_Control_Smart_Grids).

For citation purposes, cite each article independently as indicated on the article page online and as indicated below:

LastName, A.A.; LastName, B.B.; LastName, C.C. Article Title. *Journal Name* **Year**, *Article Number*, Page Range.

ISBN 978-3-03936-758-0 (Hbk)
ISBN 978-3-03936-759-7 (PDF)

Contents

About the Editors

Pierluigi Siano received an MSc degree in electronic engineering and a PhD in information and electrical engineering from the University of Salerno, Salerno, Italy, in 2001 and 2006, respectively. He is a Professor and Scientific Director of the Smart Grids and Smart Cities Laboratory with the Department of Management & Innovation Systems, University of Salerno. His research activities are centered on demand response, the integration of distributed energy resources in smart grids and the planning and management of power systems. He has co-authored more than 450 papers, including more than 250 international journal papers that received more than 8400 citations, with an h-index equal to 46. He received the Highly Cited Researcher award from the ISI Web of Science Group in 2019. He has been the Chair of the IES TC on Smart Grids. He is an Editor of the Power & Energy Society Section of *IEEE Access, IEEE Transactions on Industrial Informatics, IEEE Transactions on Industrial Electronics,* the *Open Journal of the IEEE IES and of IET Renewable Power Generation*.

Miadreza Shafie-khah received his first postdoc from the University of Beira Interior (UBI), Covilha, Portugal in 2015. He received his second postdoc from the University of Salerno, Salerno, Italy in 2016. Currently, he is a tenure-track Professor at the University of Vaasa, Vaasa, Finland. He is an Associate Editor for IET-RPG and an Editor of the IEEE OAJPE. He was considered one of the Outstanding Reviewers of the *IEEE Transactions on Sustainable Energy* in 2014 and 2017, one of the Best Reviewers of the IEEE Transactions on Smart Grid in 2016 and 2017, and one of the Outstanding Reviewers of the *IEEE Transactions on Power Systems* in 2017 and 2018. He has co-authored more than 310 papers that received more than 4500 citations, with an h-index equal to 39. He is also the editor of the book Blockchain-Based Smart Grids (Elsevier, 2020). His research interests include power market simulation, market power monitoring, power system optimization, demand response, electric vehicles, price and renewable forecasting and smart grids.

Editorial

Special Issue on Advanced Approaches, Business Models, and Novel Techniques for Management and Control of Smart Grids

Pierluigi Siano [1,*] and Miadreza Shafie-khah [2,*]

1 Department of Management & Innovation Systems, University of Salerno, 84084 Fisciano, Italy
2 School of Technology and Innovations, University of Vaasa, 65200 Vaasa, Finland
* Correspondence: psiano@unisa.it (P.S.); mshafiek@univaasa.fi (M.S.-k)

Received: 27 April 2020; Accepted: 9 May 2020; Published: 26 May 2020

1. Introduction

The current power system should be renovated to fulfill social and industrial requests and economic advances. Hence, providing economic, green, and sustainable energy are key goals of advanced societies. In order to meet these goals, the latest features of smart grid technologies need to have the potential to improve reliability, flexibility, efficiency, and resiliency. This *Special Issue* aims to encourage researchers to address the mentioned challenges.

2. Advanced Smart Grids

Considering the aim of the special issue, ten papers have been published in the area of advanced approaches and novel techniques for the management and control of smart grids, from different standpoints. In reference [1], the classification of energy management systems has been carried out from a new viewpoint. Energy management systems have been classified into four categories based on the type of storage systems. Furthermore, energy management systems have been reviewed in terms of uncertainty modeling techniques, objective functions and constraints, optimization techniques, as well as simulation and experimental results. In order to perform dynamic optimizations in real-time, new solutions are required to consider data privacy and cybersecurity. In reference [2], a paradigm has been designed to meet these requirements based on a bottom-up method to lead to misinterpretations or wrong conclusions. Fractal analysis is also employed to identify unique and independent elements of smart grids required for the design of an architectural paradigm. The processes of demand response and conservation voltage reduction are also presented. In reference [3], the greedy smallest-cost-rate path first algorithm is presented to route power from sources to loads in a digital microgrid. In such a microgrid, one distributed energy resource may supply power to one or many loads, and one or many distributed energy resources may supply the power requested by a load. Reference [3] has addressed the high complexity by using heuristics to match sources and loads and to select the smallest-cost-rate paths in the digital microgrid. Reference [4] has studied the building energy and comfort management systems. These systems have a high potential to decrease energy consumption costs as well as to enhance the efficiency of resource exploitation, by implementing strategies and policies for resource control and demand side management. In order to prevent users' disaffection and the gradual abandonment of the system, reference [4] has developed a sensor-based system for the analysis of users' habits and the early detection and prediction of user activities.

Reference [5] has studied the phase synchronization of an islanded network with large-scale distributed generations in a non-homogeneous condition. The critical coupling strength formula is reduced for different weighting cases via the synchronization condition. On this basis, three cases of Gaussian distribution, power-law distribution, and frequency-weighted distribution have been analyzed. In reference [6], an islanding detection algorithm for distributed generations has been proposed. This work presents a passive islanding detection algorithm for all types of distributed generations by combining the measured frequency, voltage, current, phase angle and the rate of changes at the point of common coupling. The proposed algorithm can detect the islanding situation, even with the exact zero power mismatches. In reference [7], a novel approach has been presented for earth fault location by using a negative sequence current. The developed method is proposed for locating single-phase earth faults on non-effectively earthed medium voltage distribution networks, and it requires only current measurements. The method is based on the analysis of negative sequence components of the currents measured at secondary substations along with medium voltage distribution feeders.

In reference [8], an augmented Prony method has been proposed for power system oscillation analysis using synchrophasor data obtained from a wide-area measurement system. Intrinsic mode functions provide an intuitive representation of the oscillatory modes, and they are mainly calculated by Hilbert–Huang transform methods. However, those methods suffer from the end effects, mode-mixing and Gibbs phenomena. To overcome the drawback, the augmented Prony method has been employed in [8] and tested on an online oscillation-monitoring framework. Another important issue for the operation of islanded microgrids is frequency stability. An inherent delay in the communication system or other parts such as sensor sampling rates may lead microgrids to have unstable operation states. Therefore, it is essential to use an optimization algorithm to determine the optimum parameters. In reference [9], the communication system delay and ultra-capacitor size along with the parameters of the secondary controller have been obtained by using a Non-dominated Sorting Genetic Algorithm II (NSGA-II). Reference [10] has proposed a black-box modeling method to identify the voltage and frequency response model of microgrids. The microgrid has been considered a two-input two-output black-box system and the simplicity of the modeling can be met by input and output ports, while the model parameters can be adjusted by the change of microgrid operating structure, which makes the model more adaptable.

3. Future Approaches, Models, and Techniques

Although the *Special Issue* has been closed, there is a need for new studies in advanced approaches, business models, and novel techniques for management and control of smart grids. The studies should address the complexity of customers' privacy, climate change, and the role of smart grids for building a greener planet.

Author Contributions: Both guest editors have equally carried out the editorial work. All authors have read and agreed to the published version of the manuscript.

Funding: This research received no external funding.

Acknowledgments: We would like to take this opportunity to record our gratitude to all the authors, hardworking and professional reviewers, and the dedicated editorial team of *Energies*.

Conflicts of Interest: The authors declare no conflict of interest.

References

1. Shayeghi, H.; Shahryari, E.; Moradzadeh, M.; Siano, P. A survey on microgrid energy management considering flexible energy sources. *Energies* **2019**, *12*, 2156. [CrossRef]
2. Ilo, A. Design of the smart grid architecture according to fractal principles and the basics of corresponding market structure. *Energies* **2019**, *12*, 4153. [CrossRef]
3. Jiang, Z.; Sahasrabudhe, V.; Mohamed, A.; Grebel, H.; Rojas-Cessa, R. Greedy algorithm for minimizing the cost of routing power on a digital microgrid. *Energies* **2019**, *12*, 3076. [CrossRef]
4. Marcello, F.; Pilloni, V.; Giusto, D. Sensor-based early activity recognition inside buildings to support energy and comfort management systems. *Energies* **2019**, *12*, 2631. [CrossRef]
5. Chen, S.; Zhou, H.; Lai, J.; Zhou, Y.; Yu, C. Phase synchronization stability of non-homogeneous low-voltage distribution networks with large-scale distributed generations. *Energies* **2020**, *13*, 1257. [CrossRef]

6. Abyaz, A.; Panahi, H.; Zamani, R.; Haes Alhelou, H.; Siano, P.; Shafie-khah, M.; Parente, M. An effective passive islanding detection algorithm for distributed generations. *Energies* **2019**, *12*, 3160. [CrossRef]
7. Farughian, A.; Kumpulainen, L.; Kauhaniemi, K. Earth fault location using negative sequence currents. *Energies* **2019**, *12*, 3759. [CrossRef]
8. Khodadadi Arpanahi, M.; Kordi, M.; Torkzadeh, R.; Haes Alhelou, H.; Siano, P. An augmented prony method for power system oscillation analysis using synchrophasor data. *Energies* **2019**, *12*, 1267. [CrossRef]
9. Alizadeh, G.A.; Rahimi, T.; Babayi Nozadian, M.H.; Padmanaban, S.; Leonowicz, Z. Improving microgrid frequency regulation based on the virtual inertia concept while considering communication system delay. *Energies* **2019**, *12*, 2016. [CrossRef]
10. Shi, Y.; Xu, D.; Su, J.; Liu, N.; Yu, H.; Xu, H. Black-box behavioral modeling of voltage and frequency response characteristic for islanded microgrid. *Energies* **2019**, *12*, 2049. [CrossRef]

Review

A Survey on Microgrid Energy Management Considering Flexible Energy Sources

Hossein Shayeghi [1,*], Elnaz Shahryari [1], Mohammad Moradzadeh [2] and Pierluigi Siano [3,*]

[1] Department of Technical Engineering, University of Mohaghegh Ardabili, 56199-11367 Ardabil, Iran; elnaz.shahryari@yahoo.com

[2] Electrical Engineering Department, Shahid Rajaee Teacher Training University, Lavizan, Trehran 16785-163, Iran; m.moradzadeh@sru.ac.ir

[3] Department of Management & Innovation Systems, University of Salerno, Via Giovanni Paolo II, 132, 84084 Fisciano (SA), Italy

* Correspondence: hshayeghi@gmail.com (H.S.); psiano@unisa.it (P.S.); Tel.: +98-4533512910 (H.S.); +39-089-96-4294 (P.S.)

Received: 4 May 2019; Accepted: 31 May 2019; Published: 5 June 2019

Abstract: Aggregation of distributed generations (DGs) along with energy storage systems (ESSs) and controllable loads near power consumers has led to the concept of microgrids. However, the uncertain nature of renewable energy sources such as wind and photovoltaic generations, market prices and loads has led to difficulties in ensuring power quality and in balancing generation and consumption. To tackle these problems, microgrids should be managed by an energy management system (EMS) that facilitates the minimization of operational costs, emissions and peak loads while satisfying the microgrid technical constraints. Over the past years, microgrids' EMS have been studied from different perspectives and have recently attracted considerable attention of researchers. To this end, in this paper a classification and a survey of EMSs has been carried out from a new point of view. EMSs have been classified into four categories based on the kind of the reserve system being used, including non-renewable, ESS, demand-side management (DSM) and hybrid systems. Moreover, using recent literature, EMSs have been reviewed in terms of uncertainty modeling techniques, objective functions (OFs) and constraints, optimization techniques, and simulation and experimental results presented in the literature.

Keywords: microgrid; energy management system; demand-side management; uncertainty; energy storage; distributed generation

---

1. Introduction

There are strong incentives to utilize distributed generations (DGs) for reducing greenhouse gases, improving power system efficiency as well as its reliability, competitive energy policies and postponement of transmission and distribution system upgrading [1]. In fact, DGs are composed of renewable units such as wind turbines (WTs), photovoltaic (PV), fuel cells (FCs), biomass along with non-renewable ones such as micro-turbines (MTs), gas engines (GEs), diesel generators (DiGs), etc. [2]. DGs eliminate the need for the transmission system by being installed near the customers [3]. Integration and control of DGs along with storage devices and flexible loads can constitute a low voltage distribution network, called a microgrid, which can be operated in isolated or grid-connected mode [4]. The generic concept of a microgrid is shown in Figure 1.

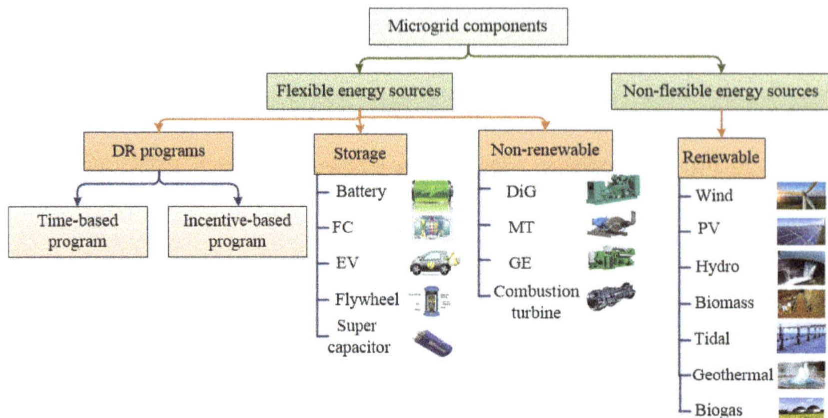

Figure 1. Microgrid components.

Microgrids often face difficulties in supplying demand due to the lack of sufficient energy generation sources. This obstacle is caused by intermittent nature of loads and renewable energy sources [5]. As a result an energy management system (EMS) is necessary to tackle this problem. EMS for a microgrid represent relatively new and popular topics that attracted lots of attention, recently. Existing review papers in literature have investigated microgrid EMSs from a different aspects, as summarized in Table 1 [6–13].

Table 1. A survey on the existing review papers of microgrid energy management system.

Ref.	Objective Function	Const.	Flexible Resources		Optimization Techniques	Microgrid Operational Mode	
			DR	ESS		Islanded	Grid-Connected
[6]	✓	✓		✓	✓	✓	
[7]	✓			✓	✓	✓	✓
[8]					✓	✓	✓
[9]				✓		✓	✓
[10]				✓	✓	✓	✓
[11]					✓	✓	✓
[12]			✓		✓	✓	
[13]	✓	✓			✓	✓	

The contributions of this review paper are:

- Microgrid EMSs have been classified into four categories based on the kind of the reserve system being used including non-renewable energy sources, energy storage system (ESS), demand-side management (DSM) and hybrid.
- Energy management modeling studies have been reviewed in terms of uncertainty modeling techniques, objective functions (OFs), constraints, and optimization techniques.
- The microgrids which are considered as the case study of different EMS papers have been reviewed in this paper.
- The scenarios of the simulation results section have been categorized.

The aim of an EMS is to determine the optimal use of DGs in order to feed the electrical loads [14]. An EMS can be operated in two modes, namely centralized and decentralize. In the centralized mode, the central controller aims to optimize the microgrid power exchanged based on the market prices and security constraints. In the decentralized mode, DGs and controllable loads have more degree of freedom [15]. As a result, the microgrid components are considered to be intelligent and try to maximize the revenue of the microgrid by communicating with each other [16]. The initial duty of EMS in both centralized and decentralized mode is to ensure the microgrid of providing load-generation balance [17]. The EMS fail in matching the generation and load, whenever the total load is higher than the maximum capacity of DGs [18] and no other additional actions are taken. A solution for this drawback is to trade power with the utility or other micro-sources, however, this solution leads to an increment of pollution, costs and the need to solve a more complex unit commitment problem as a result of the additional units [19]. Various supporting systems such as DiGs, ESSs and DSM are employed to overcome the supply-demand mismatch of a microgrids. This paper provides the literature review of microgrid EMSs by classifying the existing articles into four categories as follows:

(1) Non-renewable based EMS
(2) ESS-based EMS
(3) DSM-based EMS
(4) Hybrid systems based EMS

The remainder of the paper is organized as follows: in Section 2, the aforementioned categories are briefly introduced. A survey of uncertainty modeling techniques in EMSs is presented in Section 3. The mathematical formulation of objective functions along with constraints are presented in Section 4. The appropriate optimization techniques used by an EMS, microgrid test systems and obtained simulation results are reviewed in Sections 5–7, respectively. Finally, the conclusions are presented in Section 8.

2. Classification of EMS Literature

In this section, the necessity and utilization of the aforementioned categories are explained.

2.1. Category 1: Non-Renewable Based EMS

In case of failure or inaccessibility of ESSs, it is recommended to use non-renewable energy sources including diesel generators (DiGs), micro turbines (MTs), gas engines (GEs) or combustion turbines as a backup energy source in the microgrid. A DiG is made up of a combination of a diesel engine and an electric generator. The efficient selection of DiG depends on various factors such as load type, fuel cost, transportation cost, etc. [7].

2.2. Category 2: ESS-Based EMS

Microgrid EMSs face difficulties in the management of renewable energy sources such as wind and solar energy. This problem is due to the uncertain nature of the available energy which is caused by the difference between real-time and forecasted power production [20]. One of the solutions to tackle this problem is to utilize ESSs [21]. In most cases, ESSs maintain the power balance between generation and consumption by storing power during inexpensive or off-peak hours and discharging it during high-price or peak hours [22]. Diverse studies have been focused on the utilization of ESSs in a microgrid [23]. Utilizing ANN (Artificial neural network) for prediction of wind power generation, multi-objective energy management of microgrid considering uncertainties of wind generation in presence of ESS is studied in [24]. Karavas et al. [25] have studied the microgrid EMS considering a battery as ESS and solved the optimization problem based on distributed intelligence and MAS. Alavi et al. [26] solved the microgrid energy management problem considering a battery as a reserve energy source. In order to cover wind and solar power uncertainties, PEM has been used. Chen and Duan [27] have solved the optimal management problem of DGs with ESS in a microgrid by a MRCGA. Simultaneous capacity optimization of DGs and storage devices by considering the effects of weather condition and non-dispatchable power sources have been introduced in [28]. The EMS problem for multi-microgrid is proposed by Logenthiran et al. in [29] while the ESS being used as a reserve energy source.

2.3. Category 3: DSM-Based EMS

Another way to cope with unbalances of microgrid production and consumption is to utilize DSM programs [1]. All activities aiming to match the supply and demand by modifying time and/or shape of customers' demand profile is called DSM [30]. After the liberalization of electricity market, DSM is divided into following two categories [31]:

- Energy efficiency: which reduces consumption of the demand side by improving the efficiency of products.
- Demand response (DR): which modifies the usage of end-users in comparison to their normal consumption in response to changes in electricity price or incentive payments which aim to reduce consumption during expensive hours or when system reliability is at risk [32].

DG uncertainties and electricity price fluctuations cannot be managed and coped with for energy efficiency. As a result, special attention is given to DR programs in a microgrid and it is therefore crucial to study the impacts of various DR programs on the EMS of a microgrid.

DR programs are categorized into two groups namely: price-based and incentive-based DR programs [33]. A more detailed explanation of these categories can be found in [34].

Microgrid EMS in a deterministic case and with a multi-objective problem is studied in [35] wherein a price-quantity based DR package is utilized to manage the random nature of DGs. Falsafi et al. [36] have utilized both price-based and incentive-based DR programs to provide reserve, and to cover the wind power uncertainties in a multi-objective formulation. Nejhad et al. [37] have studied the EMS problem of microgrid as a stochastic problem in presence of DR programs. In addition, microgrid EMS taking into account effects of DR program is formulated as a MILP in [38]. A security-constrained EMS problem is considered in [39], in which frequency management is studied along with energy management and DR is utilized as a reserve energy source. The optimal operation of microgrid in the presence of electrical vehicles and responsive loads are discussed in [40] by considering wind and PV uncertainties.

2.4. Category 4: Hybrid Systems Based EMS

In this category, a combination of the abovementioned categories, two by two or all together, are surveyed to overcome microgrid EMS problems. Similar to the previous categories, this topic has also attracted lots of attention in the literature. In [41], the probabilistic coordination of DGs, ESS and DR are presented in a microgrid by performing a load reduction in the presence of security risks. Simultaneous implementation of ESS and DR is studied as a multi-objective problem by Marzband et al. [18]. Microgrid EMS in a system containing PV and wind turbines as DGs, DiG and small hydro generator as reserve power sources, DR and battery storage system is modeled by Zhao et al. [42] by using multi-agent systems (MASs). Talari et al. [43] have studied stochastic scheduling of microgrid components in presence of ESSs and DR programs. Pourmousavi et al. [44] have solved the management problem of an islanded microgrid which is composed of various DG units, storage and DR as a multi-timescale problem. The authors of [45] have utilized a pumped-storage unit and DR program in a new stochastic optimization framework to cover existed uncertainties of microgrid EMS problem. An overview of recent technologies in application of storage and DR operation of microgrids is presented in [33].

Solving the EMS problem of a microgrid is composed of various steps, including predicting uncertain parameters, modeling uncertainties, mathematical formulation of objective functions and constraints, choosing the optimization technique to solve the problem and selecting the understudying microgrid as the case study. Classification of the abovementioned steps along with sub-category of each step is shown in Figure 2. For instance, the existed uncertain parameters in microgrid EMS problem can be predicted in different time horizons such as short-term, mid-term and long-term time period. Prediction in short-term period can be performed using classical and intelligent techniques. Classical methods include ARIMA (auto-regressive integrated moving average), GHARCH (Generalized auto-regressive conditional heteroskedasticity), DR and TF (Transfer function) while the intelligent ones are ANN, FNN (fuzzy neural network) and SVR (support vector regression). Theses explanations can be expanded for other steps, too.

3. Prediction of Uncertain Parameters

Lack of information which leads to a probability of a difference between real and forecasted values is defined as uncertainty [26]. The uncertain parameters in a microgrid EMS can be generally classified into the following two categories:

- Operational parameters: These parameters include the amount of generation and load in power systems.
- Economical parameters: Economic parameters have an effect on the economic aspects of the power system and include uncertainty in fuel supply, production cost, economic growth and interest rates [46].

The prediction of uncertain parameters can be performed over various time horizons ranging from minutes to a couple of days as a short-term prediction, from several weeks to months as mid- term prediction and from multiple months to several years as long-term prediction [47]. Since the microgrid EMS problem is accomplished at hourly intervals, short-term forecasting methods are suitable to be used for this purpose. Classification of these short-term forecasting methods is represented in Figure 2 which is broken down into two categories namely classical and intelligent methods [37]. Classical techniques include well-known methods such as auto-regressive integrated moving average (ARIMA) [48], generalized auto-regressive conditional heteroskedasticity (GARCH) [49], dynamic regression (DR) and transfer function [50].

Intelligent methods are based on training historical data by mapping them on input-output of data-driven structures [37]. Artificial neural network (ANN) [51], fuzzy neural network (FNN) [52] and support vector regression (SVR) based on ARIMA can be classified in this category [53].

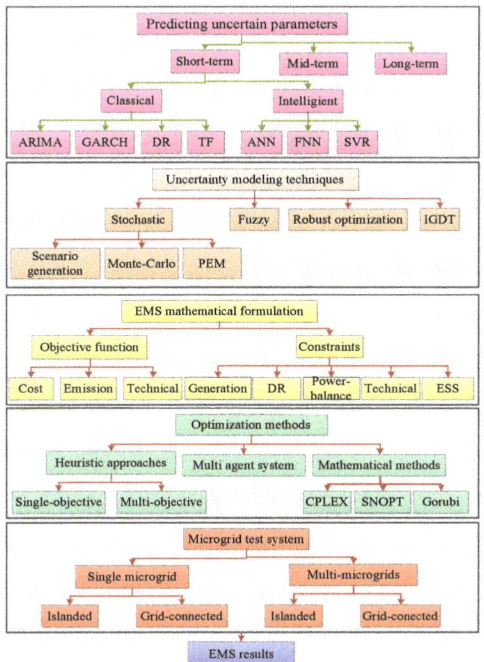

Figure 2. Overall EMS structure and underlying methods of each step.

4. Uncertainty Modeling

Uncertainty managing is one of the main difficulties decision makers have to deal with [46]. As a result, various methods have been employed in order to manage the uncertainty of the aforementioned parameters in Section 3 in microgrid energy management [46,54–57]. In this section, a review of all existing uncertainty modeling techniques used by an EMS are presented which are classified broadly into four categories as shown in Figure 2.

4.1. Stochastic Methods

Stochastic methods are used to approximate the PDF of random variables [58]. The first efforts in modeling uncertainties using stochastic methods was made by Dantzig [59]. These methods can be used, when the PDF of input parameters is known [60]. A brief introduction of three main stochastic technique are given below:

4.1.1. Scenario Generation

Modeling uncertainties through scenario generation methods needs to know the PDF. In these methods, by dividing the PDF in multiple parts, each part is assumed as a scenario with an occurrence probability proportional to the PDF value of the preferred selected section [61]. The scenario generation method is used in [40] in order to cover uncertainties related to upper, lower and expected values of wind and PV systems. As reported by Alharbi and Raahemifar [41], uncertainties of wind, solar power and load are represented by discrete probability distribution sets as scenarios. In [39], load, wind and PV power random scenarios are firstly generated using MCS and roulette wheel mechanism. Then, to improve the computational speed, a scenario reduction algorithm is employed. Various scenarios for modeling wind and PV output powers are generated by Monte Carlo simulation by Talari et al. [43].

4.1.2. MCS

In MCS-based methods, for each input parameter, a sample is generated using its PDF [62]. As MCS is a repetitive procedure, sample generation process is repeated for some iterations. Finally, histograms, statistic criteria or other approaches are used to analyze the outcome [63]. Caralis et al. [64] have used a Monte Carlo method to manage wind power uncertainties with the goal of earning profits from wind energy investment. Uncertainties related to solar radiation and load are investigated using MCS in [63] via optimization of the hybrid system.

4.1.3. Point Estimate Method (PEM)

PEM is one of the methods used to estimate the value of uncertain parameters with high computational accuracy and is based on the concept of uncertain input parameters moments [65]. This method covers uncertainties by establishing a connection between input and output variables whose main steps are explained by Hong [66]. In [26], PEM is utilized to model uncertainties of wind and solar power. Peik-Herfeh et al. [67] employed PEM to model the uncertainties of market price and generation sources via a probabilistic price-based unit commitment problem.

4.2. Fuzzy Method

Based on fuzzy theory, a degree of membership can be attributed to each uncertain parameter using membership functions. Then, an α-cut method [68] can be used to determine membership functions of output variables according to membership functions of input parameters. An effort has been made in [69] to review the application of fuzzy methods in renewable energy sources. A fuzzy method has been used by Sourodi et al. [70] to study the effects of uncertain power output of DGs on power losses in distribution networks.

4.3. Robust Optimization Method

The robust optimization method [71] was introduced to explain parameter uncertainties using uncertain boundaries. Robust optimization is suitable when there is a lack of information about PDF of parameters. This technique is used to cover load uncertainties by Alavi et al. in [26]. Robust optimization is employed for wind and load uncertainty in Ref. [72].

4.4. Information Gap Decision Theory (IGDT)

As a decision-making approach, IGDT makes minor assumptions on the structure of uncertainty [73]. This technique makes robust decisions against uncertainty of the input parameters. The optimal bidding strategy in day-ahead electricity market considering uncertainties of market price is performed by implementing IGDT in [74]. Nojavan et al. [75] proposed a method based on IGDT to evaluate procurement strategy of large consumers.

5. Mathematical Formulation of EMS

Microgrid energy management is an optimization problem which aims to properly schedule short-term operation of DGs, ESSs and controllable loads with respect to various objective functions and constraints [76]. In this section, a literature review of existing objective functions as well as constraints considered by an EMS have been elaborated. Furthermore, Figure 2 represents a classification of objective functions utilized by the EMSs along with their constraints.

5.1. Objective Functions

The EMS can manage a microgrid by solving various objective functions. Objective functions may include capital or operational costs of the microgrid. Costs related to fuel, maintenance, start-up and shut-down, degradation as well as procurement from the utility in case of power deficiency, are considered as operational costs [77]. Table 2 provides a collection of utilized EMS objective functions in literature. In this table, objective functions are reviewed from being single objective and multi-objective perspectives, too.

Table 2. Survey through collection of EMS objective functions.

Ref	OF Equation	Details	Single	Multi
[24]	$F = \sum_{t=1}^{T}\left\{\sum_{i=1}^{N_g}\left[u_i(t)P_{gi}(t)(B_{gi}(t)+K_{OM_i})+S_{gi}\lvert u_i(t-1)\rvert\right] + \sum_{j=1}^{N_{ES}}\left[u_j(t)P_{Sj}(t)B_{Sj}(t)\right]+P_{Grid}(t)B_{Grid}(t)\right\} + \sum_{t=1}^{T}\left\{(\sum_{i=1}^{T_E}\sum_{j=1}^{N} EF_{ij}P_{gi}(t))+P_{Grid}(t)EF_{grid}\right\}$	$B_{gi}(t)$ and $B_{Sj}(t)$ represent bids of *i*th DG and *j*th storage device. EF_{ij} is the emission factor of *j*th DG. In addition, $P_{gi}(t)$ and $P_{Sj}(t)$ represent power generation of *i*th DG and *j*th storage device.		✓
[28]	$COE = \frac{C_{antot}}{E_{anserved}}$	COE is the cost of energy which is computed by the ratio of total annualized cost (C_{antot}) to total annual energy served ($E_{anserved}$).	✓	
[18]	$F = \sum_{t=1}^{m}(C_t^g + Cr_t^g + C_t^{ES-} - C_t^l - C_t^{ES+} + \Omega_t) \times \Delta t$	C_t^g, Cr_t^g are the cost of energy produced by renewable and non-renewable sources. C_t^{ES-}, C_t^{ES+} represent the cost of ESS charge and discharge. C_t^l, Ω_t are the DR cost and is the penalty of the energy not supplied.	✓	

Table 2. *Cont.*

Ref	OF Equation	Details	Single	Multi
[25]	$F = NPC + \sum_{t=1}^{8760} P_b(t) + \sum_{t=1}^{8760} P_{H_2}(t) + \sum_{t=1}^{8760} P_w(t) + P_{wt} + P_{H_2T}$	NPC is the net present cost for 20 operating years. P_b, P_{H_2}, P_w, P_{wt}, P_{H_2T} are the battery, hydrogen, water, water tank and metal hydride tank penalty, respectively.	✓	
[26]	$F = CF_t^{OPR} + CF_t^{EMI} + CF_t^{RLB}$	CF_t^{OPR}, CF_t^{EMI} and CF_t^{RLB} represent the operation, emission and reliability cost of microgrid, respectively.	✓	
[27]	$F = C_{in}^{MG} + C_{op}^{MG}$ $C_{op}^{MG} = \sum_{i=1}^{L} (C_{Fi} + C_{OMi} + C_{Si} + C_{Ei}) + \sum_{j=1}^{M} C_{OMj}^{ESS} - C_{G}^{MG}$	The EMS cost composed of C_{in}^{MG} as investment cost and C_{op}^{MG} as operation cost.	✓	
[35]	$F = Cost^{Operating} + Cost^{Emission}$ $Cost^{Operating} = \sum_{t=1}^{T} (\cos t_{DG}(t) + ST_{DG}(t) + \cos t_s(t)$ $+ \cos t_{Grid}(t) + \cos t_{DR}(t))$ $Cost^{Emission} = \sum_{t=1}^{T} \{emission_{DG}(t) + emission_s(t)$ $+ emission_{Grid}(t)\}$	The objective function is considered as the operating and emission cost. $\cos t_{DG}(t)$, $ST_{DG}(t)$, $\cos t_s(t)$, $\cos t_{Grid}(t)$ and $\cos t_{DR}(t)$ represent DG cost, start-up and shut-down costs, reserve cost and cost of exchanged power with the grid, respectively.		✓
[36]	$F = F_{Cost}^{start-up} + F_{Cost}^{reserve} + F_{Cost}^{generation} + F_{Cost}^{DR} + F_{Emission}$	The objective function is composed of overall cost and emission functions.	✓	✓
[38]	$F = \sum_{t=1}^{ND} \Big\{ \sum_{a=1}^{A} [(AT_{at}.ut_{at} + (MTC_a + BT_{at}).pt_{at}).H/ND$ $+DT_a.yt_{at} + FT_a.zt_{at}] + \sum_{b=1}^{B} [((MFC_b + CF_b).pf_{bt} + \zeta_b.dpf_{bt})$ $.H/ND + EF_b.yf_{bf} + GF_b.zf_{bt}] + \sum_{c=1}^{C} [(CC_c.pdc_{ct}).H/ND]$ $+[BP_t.pgb_t - SP_t.pgs_t + CD.pde_t + CE.pex_t].H/ND\}$	MTC_a *and* MTC_b represent maintenance cost of MT and FC while CC_cexpresses the incremental cost of load shedding.	✓	
[39]	$Frequency_{MG} = \sum_{s=1}^{Ns} \pi_s (\sum_{h=1}^{Nh} \sum_{l} \lvert \Delta f(s,l,h) \rvert)$	$Frequency_{MG}$ controls microgrid frequency as the EMS OF.	✓	
[40]	$F = \omega_1 \sum_{t=1}^{T} Cost^t + \omega_2 \sum_{t=1}^{T} Q_{r,i} Emission^t$	ω_1 and ω_2 represent non-negative coefficients for adjusting objective functions while $Q_{r,i}$ is ith price penalty factor.	✓	
[41]	$F = \sum_{s \in S} \lambda_s [\sum_{k \in K} \sum_{j \in J} (C_j(P_{j,k,s}) + SU_{j,k}) + \sum_{k \in K} C_{ES}.(V_{k,s}^{CH} + V_{k,s}^{DCH})$ $+ \sum_{k \in K} P_{k,s}^{Int,R-C-I}.C_k^{Int,R-C-I} + \sum_{k \in K} \Delta P_{k,s}^{do,R-C-I}.C_k^{DR,R-C-I}]$	The first term represents operating cost of DGs while the latter one is the operating cost of ESS. The expected cost of power interruption and responsive demand are computed by third and last terms, respectively.	✓	
[43]	$F = \sum_{t=1}^{T} \Big\{ \sum_{n=1}^{N} (P_{n,t}B_{n,t} + SU_n \times y_{n,t} + SD_n \times z_{n,t} + c\pi_{n,t}^{U} SR_{n,t}^{U}$ $+ c\pi_{n,t}^{D} SR_{n,t}^{D}) + \sum_{d=1}^{ND} CDR_{d,t} + \sum_{s=1}^{S} \Pr_{t,s} SC_{t,s}\}$	The cost function is composed of generation, trade-off, start-up and shut down costs of DGs as well as up and down reserves of demand response and security cost of the network.	✓	
[78]	$F = \sum_{t \in T} C_{t,money} + \sum_{t \in T} C_{t,money}^{startup} - \sum_{t \in T} P_{t,money} + \sum_{t \in T} \sum_{t \in T} \mu_{t,g}.\pi_g$	$C_{t,money}$, $C_{t,money}^{startup}$ denote the operation and start-up costs while $P_{t,money}$ is the total revenue. The last term represents penalty of the unmet load.	✓	
[79]	$F = \sum_{k_t} [(\sum_{g} a_g P_{g,k_t}^2 + b_g P_{g,k_t} + c_g w_{g,k_t} + C_{\sup} u_{g,k_t} + C_{sdn} v_{g,k_t})$ $+ d_s P_{shed,k_t} + d_c P_{curt,k_t}] \Delta t_{k_t}$	P_{g,k_t} is the generated power by DGs while P_{shed,k_t} and P_{curt,k_t} denote the amount of shedded and curtailed load, respectively.	✓	
[80,81]	$J = \sum J_{ij} \times x_{ij} \quad for\ 1 \le i \le j \le N_n$	J is the cost of transmission network, J_{ij} denotes the cost of interconnecting nodes.	✓	

5.2. Constraints

Different constraints can have an effect on energy management of a microgrid. For instance, maximum and minimum limits of power generation units must be satisfied to secure their safe and economic performance [82]. The balance between generation and consumption is another necessity of the system. The charge and discharge rates of ESS are also constrained. Violation of these constraints can lead to damaging effects on lifetime and efficiency of the ESS. Technical constraints of a microgrid include voltage at buses, feeder currents, frequency security aspects, start-up and shut-down reserve constraints, as well as ramping limits. In some of the studies which additionally consider responsive loads, constraints related to DR program must be satisfied. Table 3 provides a summary of the considered constraints used in the formulation of microgrid EMS.

Table 3. Survey through collection of EMS constraints.

Ref	Power Balance	Generation									ESS	DR	Technical
		DiG	GE	Biomass	MT	Wind	PV	FC	EV	Grid			
[24]	✓				✓	✓	✓	✓		✓	✓		
[28]	✓			✓		✓	✓				✓	✓	✓
[18]	✓				✓	✓	✓				✓	✓	
[25]	✓					✓	✓	✓			✓	✓	
[26]	✓	✓			✓	✓	✓				✓		✓
[27]	✓				✓		✓	✓		✓	✓		
[35]	✓				✓	✓	✓	✓		✓	✓	✓	
[36]	✓					✓						✓	
[38]	✓				✓			✓				✓	
[39]	✓		✓		✓			✓				✓	✓
[40]	✓				✓	✓	✓	✓	✓	✓		✓	
[41]	✓					✓	✓				✓	✓	
[43]	✓				✓	✓	✓	✓			✓	✓	✓
[78]	✓			✓		✓	✓				✓		✓
[79]	✓	✓			✓	✓					✓	✓	

6. Optimization Techniques of a Microgrid Performed by an EMS

The literature review reveals that various techniques have been utilized by researchers in order to solve the aforementioned optimization problems [17]. In this paper, these methods have been classified as follows:

6.1. Heuristic Approaches

Heuristic optimization techniques use exploratory approaches to solve the optimization problems while are unable to assure optimality of the obtained results [6]. A deterministic energy management problem is solved via multi-period gravitational search algorithm (MGSA) by Marzband and Ghadimi in [18]. In [35], the EMS optimization problem has been considered as a continuous multi-objective problem and is solved using multi-objective particle swarm optimization (MPSO) algorithm while it is solved by single-objective PSO in [26,40]. Motevasel and Seifi [24] have employed multi-objective Multi-objective Bacterial Foraging Optimization (MBFO) to solve complex and large-scale EMS problem. To deal with non-smooth cost functions, authors in [27,79] employed MRCGA. Implementation of multi-period ABC optimization algorithm to obtain an economic schedule of ESS and controllable loads of an islanded microgrid is studied in [83].

6.2. MAS

MAS is a smart system which is composed of multiple intelligent agents that are connected to each other within an environment [15]. The goal of MAS is to solve the optimization problems which are too complicated for a single agent. By assuming each member of microgrid as an agent, MAS has been used in [42] to find the optimal solution by the EMS problem. Optimal results of decentralized EMS problem have been studied using MAS approach by Karavas et al. [25]. A three-stage algorithm based on MAS is presented by Logenthiran et al. [29] to model the EMS problem in a multi-microgrid environment in which first stage schedules each microgrid to satisfy its load. The second and third stages determine microgrid bids and export power bids, respectively. Authors of Ref. [84] have utilized trust and reputation models to interconnect MAS effectively. In [85], a novel agent-based micro-storage management technique has been presented that allows all (individually-owned) storage devices in the system to converge a profitable and efficient behavior. A class of MAS technique is Distributed Constraint Optimization Problems (DCOP) in which a group of agents try to minimize the total cost related to a set of constraints. In a DCOP formulation the decision problem is translated to a set of constraints [86]. Long-term scheduling of microgrid is solved as an optimization problem using distributed constraint optimization (DCOP) by authors of [87]. Application of DCOP to solve large problems has difficulty as each agent can handle a single variable. So, a new multi-variable agent (MVA) DCOP decomposition method has been presented by athors of [88].

6.3. *Mathematical Methods*

Different kinds of software exist in order to solve microgrid EMS problems. Some of these solvers are mentioned below:

6.3.1. CPLEX Solver

CPLEX is a solver of GAMS package which can solve integer and linear problems [6]. In [39], the EMS problem is modeled as a MILP considering uncertainties and scenario generation process, and has been solved by the CPLEX solver of the GAMS environment. The multi-scenario MILP model has been utilized to mathematically formulate the EMS problem in [41], and is solved by CPLEX solver. CPLEX solver has been utilized to solve MIP scheduling problem of microgrid by Talari et al. [43]. A large-scale MILP problem in [36] is solved also by CPLEX solver. Another EMS problem is modeled as a MILP in GAMS software and solved via CPLEX solver [79].

6.3.2. SNOPT Solver

SNOPT is another software package of GAMS which is capable of solving nonlinear optimization problems [45]. The EMS mathematical formulation in [28] is a nonlinear programming problem, and has been solved by GAMS SNOPT solver.

6.3.3. Gurobi Optimizer

Gurobi optimizer is used to solve MILP problems [89]. In [78], the EMS problem is formulated as a MILP problem and has been solved using Gurobi optimizer. Tenfen and Finardi [38] have employed the same solver for computational purposes.

7. Microgrid Test Systems

For the sake of evaluating the performance of the aforementioned EMS algorithms, they have been applied on the majority of microgrids. There are some efforts to sum up the studied microgrid test systems in the area of EMS in the literature [20]. Reference [90] has proposed the classification of microgrids based on their topologies. A review of existing real-world microgrid test systems worldwide is presented in [20,91]. Evaluation of microgrids which are applied in real-life as well as laboratory test systems are surveyed in [92]. Nonetheless, in this section we will solely focus on the test systems which are exploited in EMS studies. Table 4 has classified these test systems from various panoramas including single-microgrid, multi-microgrid, islanded and grid-connected systems.

Table 4. Review of understudying microgrid test systems.

	Ref	Microgrid Mode		Energy Source			Schematic Diagram of Microgrid
		Islanded	Grid-connected	Type	Min power (kW)	Max power (kW)	
Single Microgrid	[24]		✓	WT	0	20	
				PV	0	25	
				MT	6	30	
				FC	3	30	
				ESS	−30	30	
				Grid	−90	90	

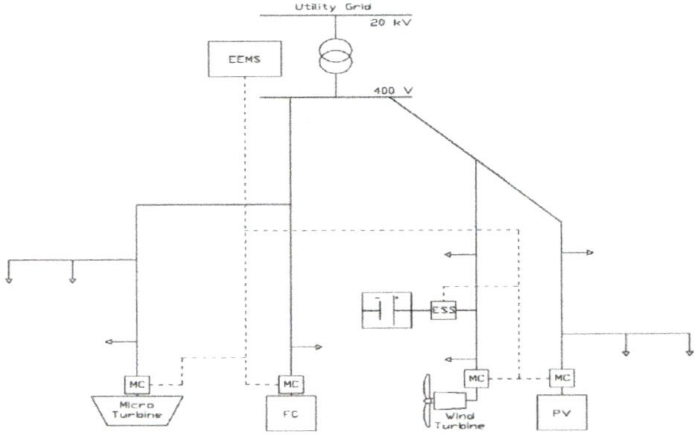

Table 4. *Cont.*

[28]	✓	WT PV Biomass Battery	0 0 0 −4500	250 250 1100 6500
[26]	✓	WT PV MT DiG Battery	0 0 12.74 11.11 30	250 200 75 60 90

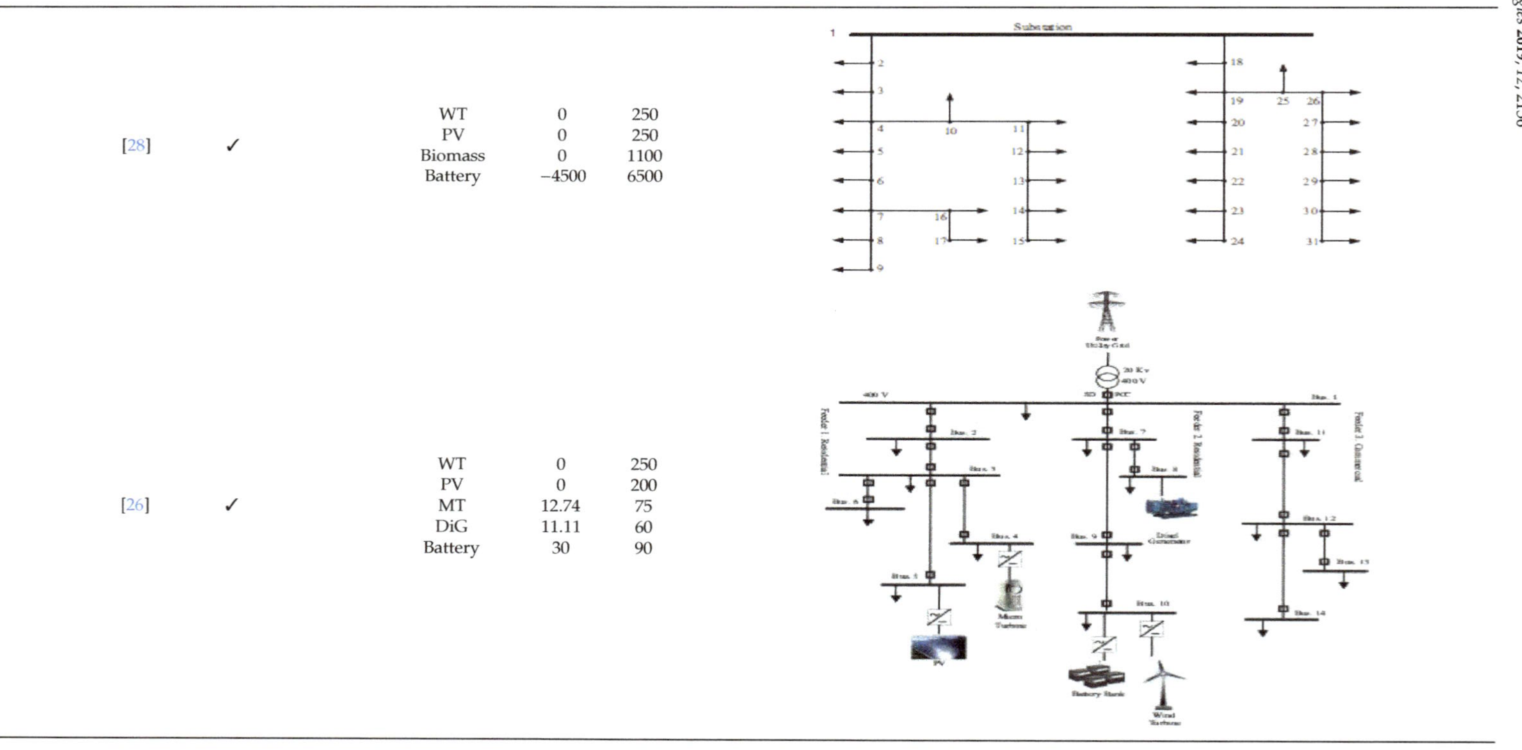

Table 4. *Cont.*

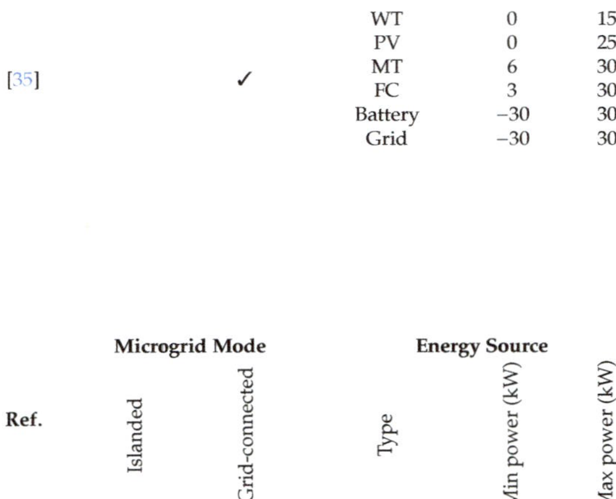

Ref.	Microgrid Mode: Islanded	Microgrid Mode: Grid-connected	Energy Source: Type	Energy Source: Min power (kW)	Energy Source: Max power (kW)	Schematic Diagram of Microgrid
[35]		✓	WT PV MT FC Battery Grid	0 0 6 3 −30 −30	15 25 30 30 30 30	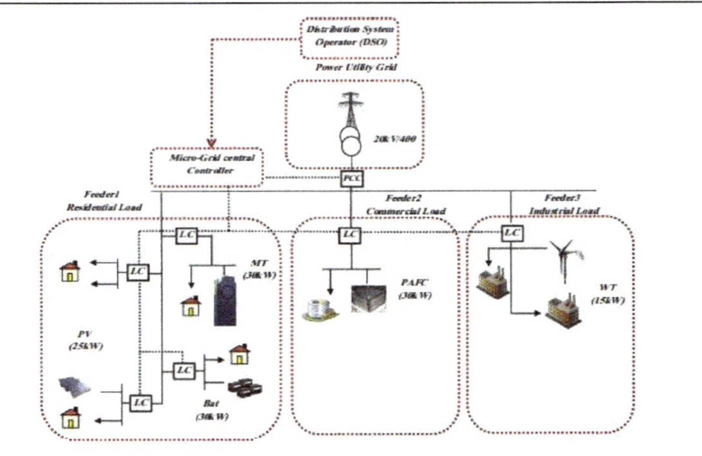

Table 4. *Cont.*

[38]		✓	WT PV MT FC Battery Grid	0 0 6 7 3 −15	7 4.5 30 10 15 15	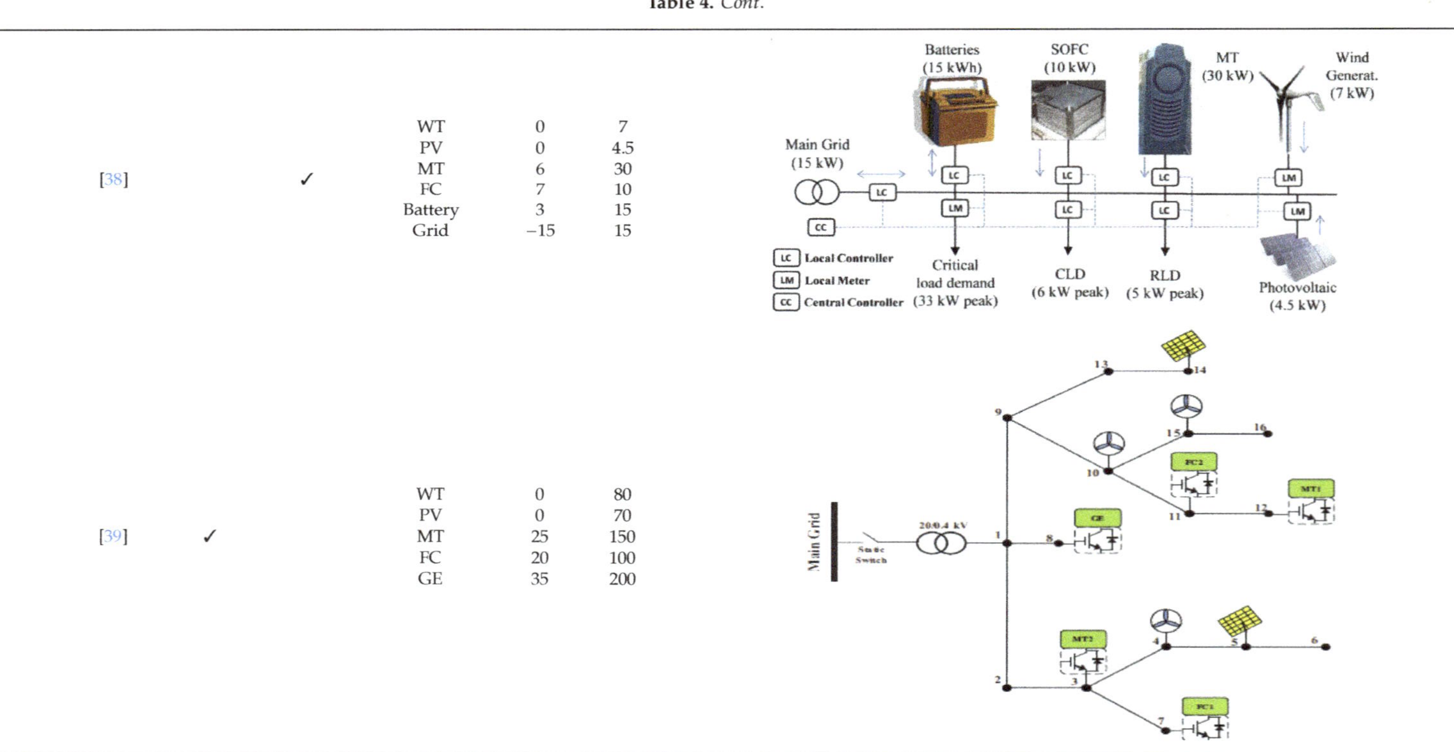
[39]	✓		WT PV MT FC GE	0 0 25 20 35	80 70 150 100 200	

Table 4. *Cont.*

[40]	✓ 	WT	0	15	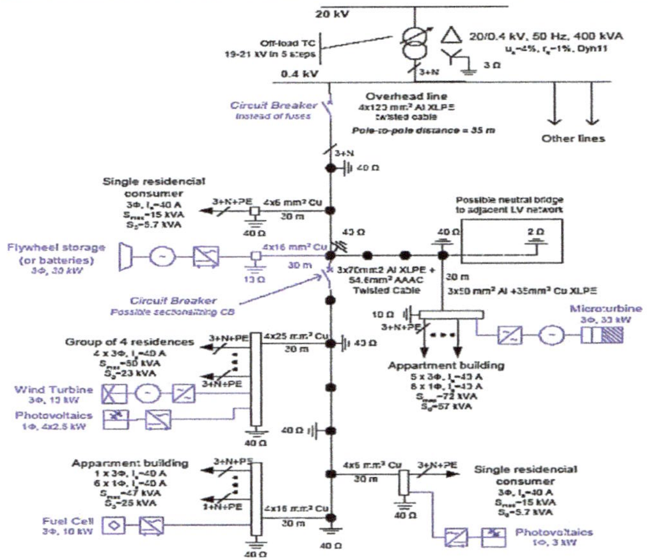
		PV	0	30	
		MT	6	30	
		FC	3	30	
		DiG	2	50	
		EV	0	1.5	

Table 4. *Cont.*

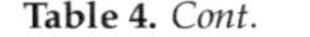

Ref.	Microgrid Mode		Energy Source			Schematic Diagram of Microgrid
	Islanded	Grid-connected	Type	Min power (kW)	Max power (kW)	
[93]	✓	✓	WT PV MT DiG FC Battery	0 0 12.74 11.11 14 30	250 200 75 60 80 90	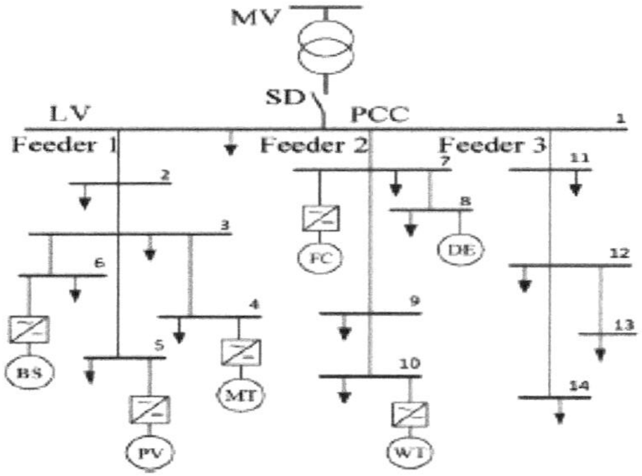

Table 4. *Cont.*

	[94]	✓		WT PV FC MT	0 0 3 6	15 2.5 30 30
Multi-Microgrids	[29]	✓	A B C	Thermal PV WT Battery	10 0 0 0	45 360 140 500

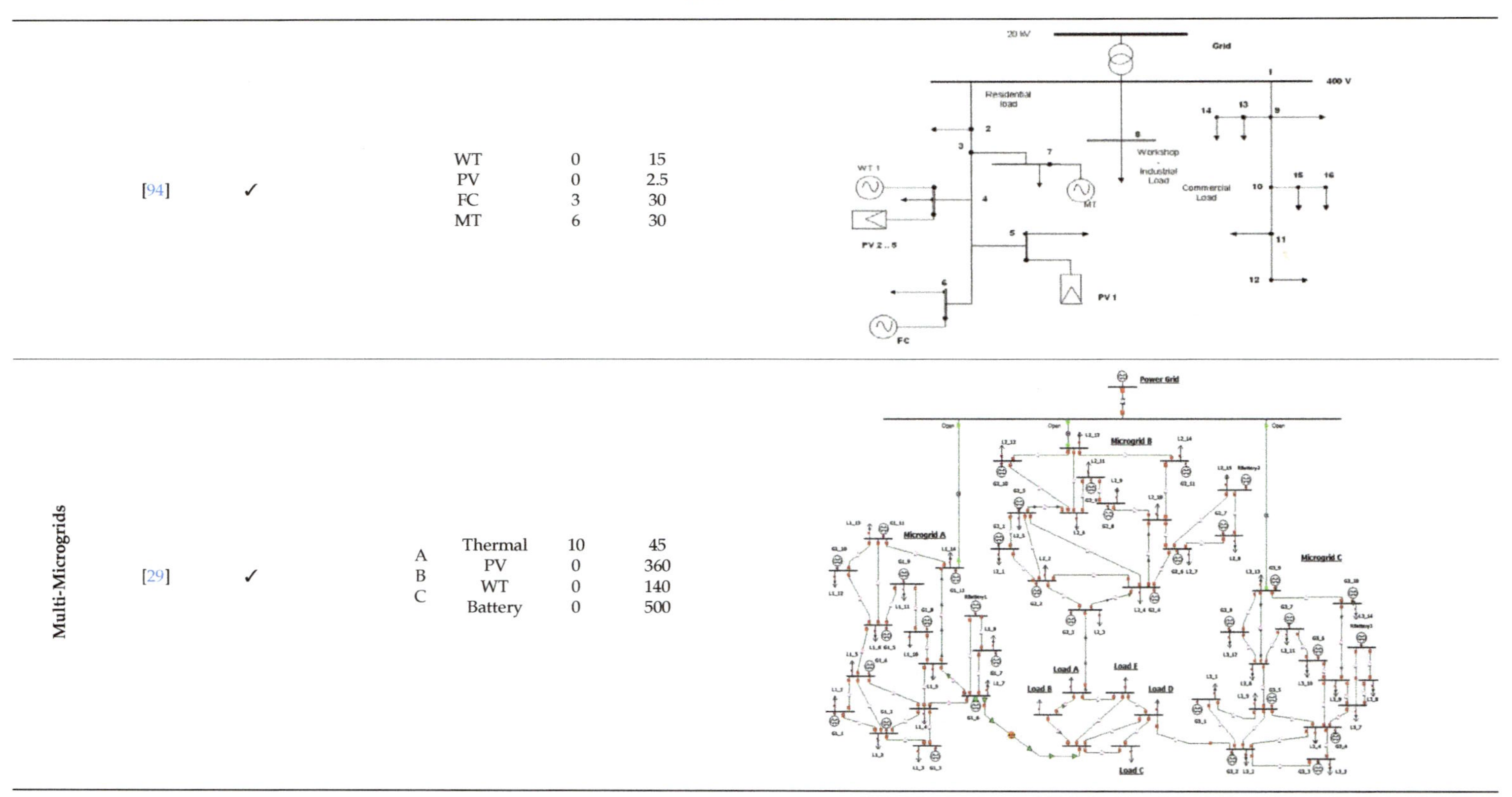

Table 4. *Cont.*

[95]	✓	1	$DG_{1,1}$ $DG_{1,2}$	0 0	5000 4000	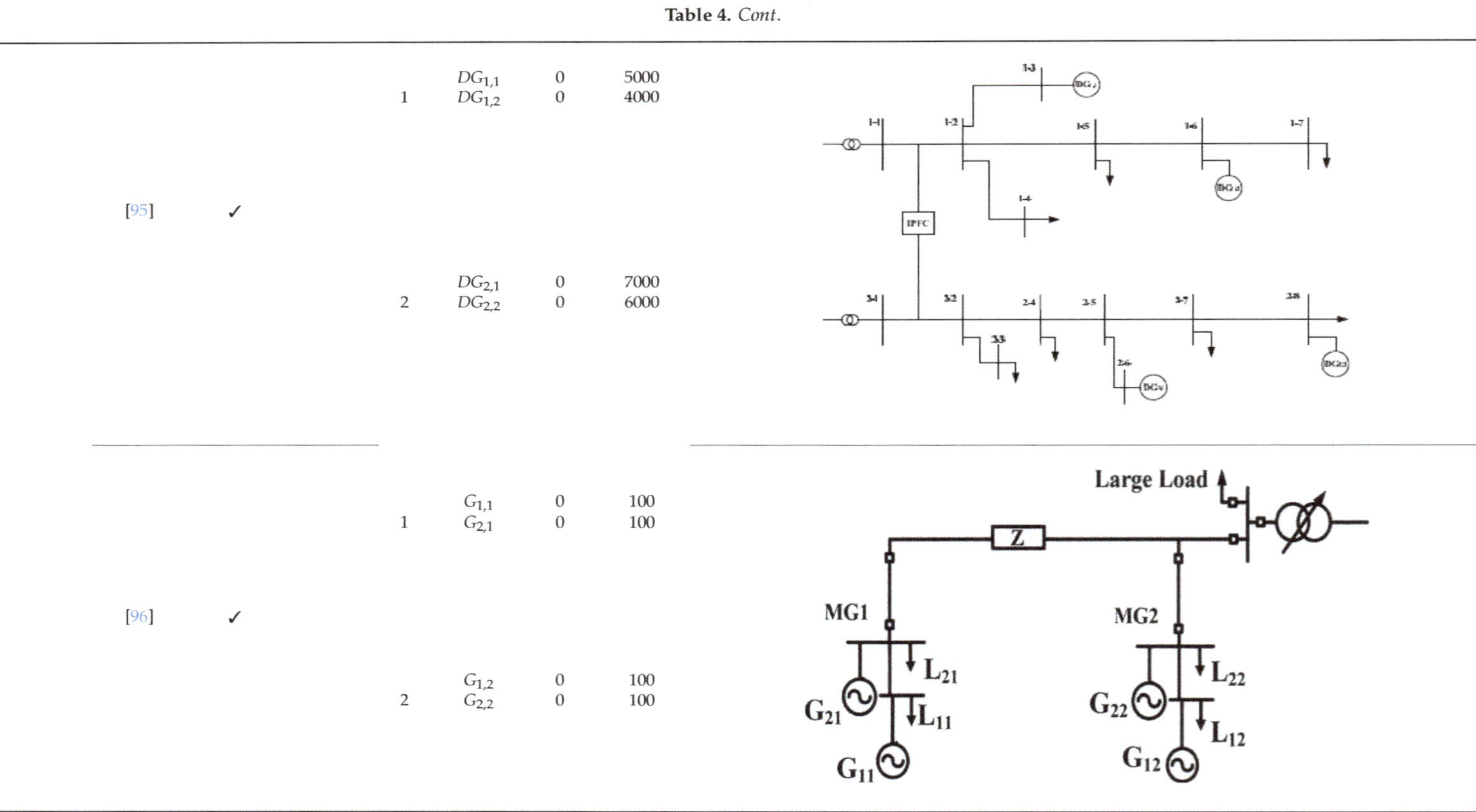
		2	$DG_{2,1}$ $DG_{2,2}$	0 0	7000 6000	
[96]	✓	1	$G_{1,1}$ $G_{2,1}$	0 0	100 100	
		2	$G_{1,2}$ $G_{2,2}$	0 0	100 100	

Table 4. *Cont.*

			WT	0	60	
		1	PV	0	60	
			MT	0	70	
		2	MT	0	30	
		3	WT	0	50	
			MT	0	60	
[97]	✓					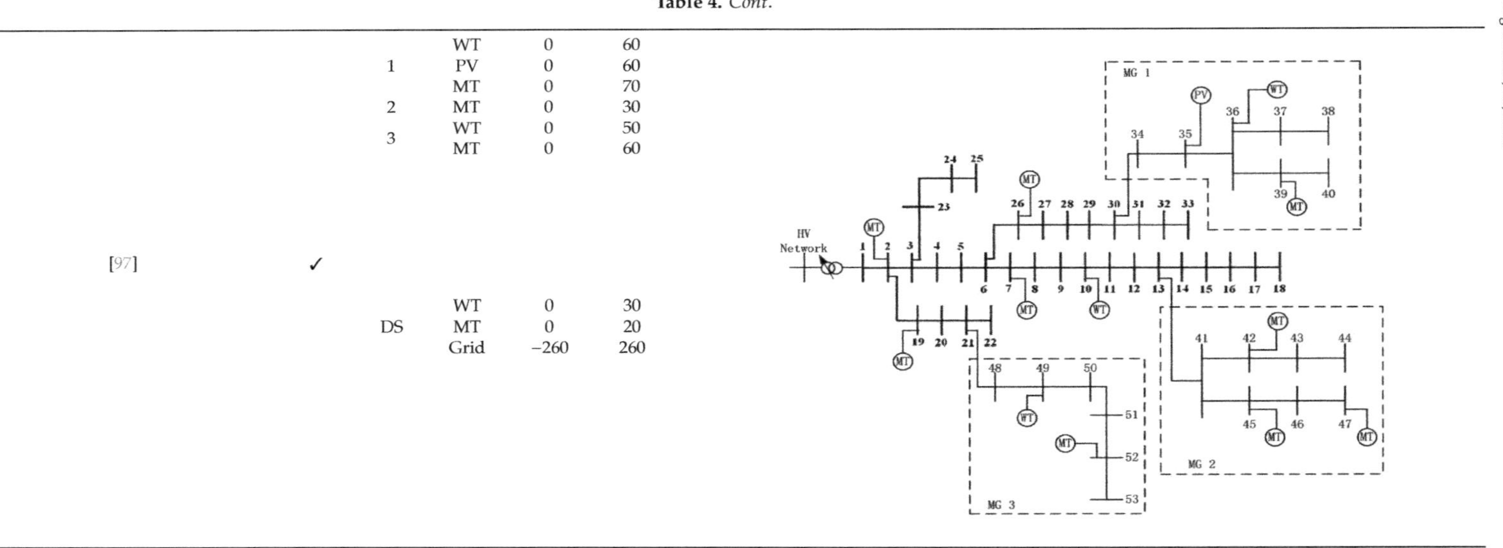
			WT	0	30	
		DS	MT	0	20	
			Grid	−260	260	

8. EMS Simulation Scenarios

The aforementioned EMS approaches are implemented on typical microgrid test systems in order to optimize the corresponding objective functions while considering various constraints in a 24-h time interval. Table 5 briefly summarizes the obtained simulation scenarios of reviewed papers to help researchers in directing their research efficiently. According to the first row of Table 5, three scenarios are considered in the results and simulation section of [24]. In the first scenario, all components of microgrid are operating within their limits and limited trade is available with the main grid. However, during the second scenario, WT operates at its maximum capacity while the third scenario studies effect of unlimited power exchange between microgrid and the upper utility. Same as the first row, remain rows of Table 5 explain the number and explanation corresponding to the simulation scenario of each reference.

Table 5. Scenarios of EMS optimization problem results.

Ref	Scenario Number	Details
[24]	1	All units operate within their limits and microgrid has limited trades with upper utility.
	2	WT operates at its maximum capacity while the rest are same as scenario 1.
	3	Unlimited energy is exchanged between microgrid and upper utility.
[28]	1	Microgrid losses are compared in islanded and grid-connected mode.
	2	The cost of energy for grid-connected and islanded microgrid are compared.
[18]	1	Normal operation.
	2	Sudden load increasing.
	3	Plug and play ability.
[26]	1	Microgrid supplies whole demand from utility grid while DGs and ESSs are neglected.
	2	DGs and ESSs have added to the microgrid while uncertain parameters are deterministic.
	3	Generated power of WT and PV are probabilistic while the load is deterministic.
	4	All of the uncertain parameters are probabilistic.
[27]	1,2	Microgrid can absorb power from utility grid in absence of ESS while the capacity of DG sources are fixed and variable.
	3,4	Microgrid can absorb power from utility grid in presence of ESS while the capacity of DG sources are fixed and variable.
	5,6	Microgrid exchanges power by utility grid in absence of ESS while the capacity of DG sources are fixed and variable.
	7,8	Microgrid exchanges power by utility grid in presence of ESS while the capacity of DG sources are fixed and variable.
[35]	1,2	Operating cost is minimized without and with DR program.
	3,4	Pollutant emission is minimized without and with DR program.
	5,6	Operating cost and pollutant emission are minimized simultaneously, without and with DR program.
[38]	1	The base case with the existing data.
	2	Extreme case by multiplying load demands by 2.
	3	Extreme case by multiplying PV and WT generation by 4 and 5.
	4	Base case along with classical modeling of MT and FC.
[39]	1	Microgrid EMS without DR.
	2	Microgrid EMS with DR.
[40]	1	Microgrid EMS without DR and EV.
	2	Microgrid EMS with DR but without EV.
	3	Microgrid EMS with DR and EV.
[41]	1	Base case without ESS and DR to focus on the generation and load uncertainty.
	2	Addition of ESS to scenario 1.
	3	Addition of DR to scenario 1.
	4	Addition of ESS and DR to scenario 1.
[43]	1	35% of peak load proportional to each hour is considered as deterministic reserve.
	2	Stochastic management of microgrid is performed without DR.
	3	DR is added to scenario 2.

9. Conclusions

This review paper has summarized energy management approaches for microgrids from a new perspective and classified it into four categories, namely non-renewable, ESS, DSM and hybrid- based EMS. This paper also provides a compendium on the modeling uncertainties associated with microgrid EMS, objective functions and constraints of microgrid formulation as well as many tools and techniques for solving this optimization problem. It is worth mentioning that considering various objective functions along with different technical and economic constraints has a large effect on the obtained EMS results. A brief review of test systems including single-microgrid and multi-microgrids is also described in this paper. A number of EMS simulation scenarios has been included while each one is an implemented simulation state. Furthermore, this work will help in identifying poorly researched areas in microgrid EMS for further investigation.

Author Contributions: H.S. supervised writing the paper, E.S wrote the paper, M.M. and P.S. checked and proofread the manuscript.

Funding: "This research received no external funding"

Conflicts of Interest: "The authors declare no conflict of interest."

Abbreviations

DG	Distributed Generation	ANN	Artificial Neural Network
ESS	Energy Storage System	FNN	Fuzzy Neural Network
EMS	Energy Management System	SVR	Support Vector Regression
DR	Demand Response	PDF	Probability Distribution Function
WT	Wind Turbine	PEM	Point Estimate Method
PV	Photo-Voltaic	MF	Membership Function
FC	Fuel Cell	IGDT	Information Gap Decision Theory
MT	Micro Turbine	MRCGA	Matrix Real-coded Genetic Algorithm
GE	Gas Engine	MAS	Multi-agent System
DiG	Diesel Generator	MILP	Mixed Integer Linear Programing
DSM	Demand-side Management	GARCH	Generalized Auto-regressive Conditional Heteroskedasticity
DR	Demand Response	ARIMA	Auto-regressive Integrated Moving Average
MCS	Monte Carlo Simulation	PEM	Point Estimate Method
MGSA	Multi-period Gravitational Search Algorithm	MPSO	Multi-objective Particle Swarm Optimization
MBFO	Multi-objective Bacterial Foraging Optimization	ABCO	Artificial Bee Colony
PSO	Particle Swarm Optimization	DCOP	Distributed Constraint Optimization Problems

References

1. Wang, D.; Qiu, J.; Reedman, L.; Meng, K.; Lai, L.L. Two-stage energy management for networked microgrids with high renewable penetration. *Appl. Energy* **2018**, *226*, 39–48. [CrossRef]
2. kianmehr, E.; Nikkhah, S.; Rabiee, A. Multi-objective stochastic model for joint optimal allocation of DG units and network reconfiguration from DG owner's and DisCo's perspectives. *Renew. Energy* **2019**, *132*, 471–485. [CrossRef]
3. Aboli, R.; Ramezani, M.; Falaghi, H. Joint optimization of day-ahead and uncertain near real-time operation of microgrids. *Int. J. Electr. Power Energy Syst.* **2019**, *107*, 34–46. [CrossRef]
4. Sedighizadeh, M.; Esmaili, M.; Jamshidi, A.; Ghaderi, M.-H. Stochastic multi-objective economic-environmental energy and reserve scheduling of microgrids considering battery energy storage system. *Int. J. Electr. Power Energy Syst.* **2019**, *106*, 1–16. [CrossRef]

5. Zheng, Y.; Jenkins, B.M.; Kornbluth, K.; Kendall, A.; Træholt, C. Optimization of a biomass-integrated renewable energy microgrid with demand side management under uncertainty. *Appl. Energy* **2018**, *230*, 836–844. [CrossRef]
6. Ahmad Khan, A.; Naeem, M.; Iqbal, M.; Qaisar, S.; Anpalagan, A. A compendium of optimization objectives, constraints, tools and algorithms for energy management in microgrids. *Renew. Sustain. Energy Rev.* **2016**, *58*, 1664–1683. [CrossRef]
7. Fathima, A.H.; Palanisamy, K. Optimization in microgrids with hybrid energy systems—A review. *Renew. Sustain. Energy Rev.* **2015**, *45*, 431–446. [CrossRef]
8. Meng, L.; Sanseverino, E.R.; Luna, A.; Dragicevic, T.; Vasquez, J.C.; Guerrero, J.M. Microgrid supervisory controllers and energy management systems: A literature review. *Renew. Sustain. Energy Rev.* **2016**, *60*, 1263–1273. [CrossRef]
9. Palizban, O.; Kauhaniemi, K.; Guerrero, J.M. Microgrids in active network management—Part I: Hierarchical control, energy storage, virtual power plants, and market participation. *Renew. Sustain. Energy Rev.* **2014**, *36*, 428–439. [CrossRef]
10. Gamarra, C.; Guerrero, J.M. Computational optimization techniques applied to microgrids planning: A review. *Renew. Sustain. Energy Rev.* **2015**, *48*, 413–424. [CrossRef]
11. Rahman, H.A.; Majid, M.S.; Jordehi, A.R.; Kim, G.C.; Hassan, M.Y.; Fadhl, S.O. Operation and control strategies of integrated distributed energy resources: A review. *Renew. Sustain. Energy Rev.* **2015**, *51*, 1412–1420. [CrossRef]
12. Vardakas, J.S.; Zorba, N.; Verikoukis, C.V. A Survey on Demand Response Programs in Smart Grids: Pricing Methods and Optimization Algorithms. *IEEE Commun. Surv. Tutor.* **2015**, *17*, 152–178. [CrossRef]
13. Korolko, N.; Sahinoglu, Z. Robust Optimization of EV Charging Schedules in Unregulated Electricity Markets. *IEEE Trans. Smart Grid* **2017**, *8*, 149–157. [CrossRef]
14. Li, B.; Roche, R.; Paire, D.; Miraoui, A. A price decision approach for multiple multi-energy-supply microgrids considering demand response. *Energy* **2019**, *167*, 117–135. [CrossRef]
15. Dou, C.; Lv, M.; Zhao, T.; Ji, Y.; Li, H. Decentralised coordinated control of microgrid based on multi-agent system. *IET Gener. Transm. Distrib.* **2015**, *9*, 2474–2484. [CrossRef]
16. Katiraei, F.; Iravani, R.; Hatziargyriou, N.; Dimeas, A. Microgrids management. *IEEE Power Energy Mag.* **2008**, *6*, 54–65. [CrossRef]
17. Theo, W.L.; Lim, J.S.; Ho, W.S.; Hashim, H.; Lee, C.T. Review of distributed generation (DG) system planning and optimisation techniques: Comparison of numerical and mathematical modelling methods. *Renew. Sustain. Energy Rev.* **2017**, *67*, 531–573. [CrossRef]
18. Marzband, M.; Ghadimi, M.; Sumper, A.; Domínguez-García, J.L. Experimental validation of a real-time energy management system using multi-period gravitational search algorithm for microgrids in islanded mode. *Appl. Energy* **2014**, *128*, 164–174. [CrossRef]
19. Shokri Gazafroudi, A.; Afshar, K.; Bigdeli, N. Assessing the operating reserves and costs with considering customer choice and wind power uncertainty in pool-based power market. *Int. J. Electr. Power Energy Syst.* **2015**, *67*, 202–215. [CrossRef]
20. Lidula, N.W.A.; Rajapakse, A.D. Microgrids research: A review of experimental microgrids and test systems. *Renew. Sustain. Energy Rev.* **2011**, *15*, 186–202. [CrossRef]
21. Pandit, M.; Srivastava, L.; Sharma, M. Environmental economic dispatch in multi-area power system employing improved differential evolution with fuzzy selection. *Appl. Soft Comput.* **2015**, *28*, 498–510. [CrossRef]
22. Garcia-Gonzalez, J.; Muela, R.M.R.; dl Santos, L.M.; Gonzalez, A.M. Stochastic Joint Optimization of Wind Generation and Pumped-Storage Units in an Electricity Market. *IEEE Trans. Power Syst.* **2008**, *23*, 460–468. [CrossRef]
23. Guo, L.; Liu, W.; Li, X.; Liu, Y.; Jiao, B.; Wang, W.; Wang, C.; Li, F. Energy Management System for Stand-Alone Wind-Powered-Desalination Microgrid. *IEEE Trans. Smart Grid* **2016**, *7*, 1079–1087. [CrossRef]
24. Motevasel, M.; Seifi, A.R. Expert energy management of a micro-grid considering wind energy uncertainty. *Energy Convers. Manag.* **2014**, *83*, 58–72. [CrossRef]
25. Karavas, C.-S.; Kyriakarakos, G.; Arvanitis, K.G.; Papadakis, G. A multi-agent decentralized energy management system based on distributed intelligence for the design and control of autonomous polygeneration microgrids. *Energy Convers. Manag.* **2015**, *103*, 166–179. [CrossRef]

26. Alavi, S.A.; Ahmadian, A.; Aliakbar-Golkar, M. Optimal probabilistic energy management in a typical micro-grid based-on robust optimization and point estimate method. *Energy Convers. Manag.* **2015**, *95*, 314–325. [CrossRef]
27. Chen, C.; Duan, S. Optimal allocation of distributed generation and energy storage system in microgrids. *IET Renew. Power Gener.* **2014**, *8*, 581–589. [CrossRef]
28. Sfikas, E.E.; Katsigiannis, Y.A.; Georgilakis, P.S. Simultaneous capacity optimization of distributed generation and storage in medium voltage microgrids. *Int. J. Electr. Power Energy Syst.* **2015**, *67*, 101–113. [CrossRef]
29. Logenthiran, T.; Srinivasan, D.; Khambadkone, A.M. Multi-agent system for energy resource scheduling of integrated microgrids in a distributed system. *Electr. Power Syst. Res.* **2011**, *81*, 138–148. [CrossRef]
30. Ramin, D.; Spinelli, S.; Brusaferri, A. Demand-side management via optimal production scheduling in power-intensive industries: The case of metal casting process. *Appl. Energy* **2018**, *225*, 622–636. [CrossRef]
31. Behrangrad, M. A review of demand side management business models in the electricity market. *Renew. Sustain. Energy Rev.* **2015**, *47*, 270–283. [CrossRef]
32. Shayeghi, H.; Sobhani, B. Integrated offering strategy for profit enhancement of distributed resources and demand response in microgrids considering system uncertainties. *Energy Convers. Manag.* **2014**, *87*, 765–777. [CrossRef]
33. Shahryari, E.; Shayeghi, H.; Mohammadi-ivatloo, B.; Moradzadeh, M. An improved incentive-based demand response program in day-ahead and intra-day electricity markets. *Energy* **2018**, *155*, 205–214. [CrossRef]
34. Colson, C.M.; Nehrir, M.H. Comprehensive Real-Time Microgrid Power Management and Control with Distributed Agents. *IEEE Trans. Smart Grid* **2013**, *4*, 617–627. [CrossRef]
35. Aghajani, G.R.; Shayanfar, H.A.; Shayeghi, H. Presenting a multi-objective generation scheduling model for pricing demand response rate in micro-grid energy management. *Energy Convers. Manag.* **2015**, *106*, 308–321. [CrossRef]
36. Falsafi, H.; Zakariazadeh, A.; Jadid, S. The role of demand response in single and multi-objective wind-thermal generation scheduling: A stochastic programming. *Energy* **2014**, *64*, 853–867. [CrossRef]
37. Shayeghi, H.; Ghasemi, A.; Moradzadeh, M.; Nooshyar, M. Simultaneous day-ahead forecasting of electricity price and load in smart grids. *Energy Convers. Manag.* **2015**, *95*, 371–384. [CrossRef]
38. Tenfen, D.; Finardi, E.C. A mixed integer linear programming model for the energy management problem of microgrids. *Electr. Power Syst. Res.* **2015**, *122*, 19–28. [CrossRef]
39. Rezaei, N.; Kalantar, M. Stochastic frequency-security constrained energy and reserve management of an inverter interfaced islanded microgrid considering demand response programs. *Int. J. Electr. Power Energy Syst.* **2015**, *69*, 273–286. [CrossRef]
40. Rabiee, A.; Sadeghi, M.; Aghaeic, J.; Heidari, A. Optimal operation of microgrids through simultaneous scheduling of electrical vehicles and responsive loads considering wind and PV units uncertainties. *Renew. Sustain. Energy Rev.* **2016**, *57*, 721–739. [CrossRef]
41. Alharbi, W.; Raahemifar, K. Probabilistic coordination of microgrid energy resources operation considering uncertainties. *Electr. Power Syst. Res.* **2015**, *128*, 1–10. [CrossRef]
42. Zhao, B.; Xue, M.; Zhang, X.; Wang, C.; Zhao, J. An MAS based energy management system for a stand-alone microgrid at high altitude. *Appl. Energy* **2015**, *143*, 251–261. [CrossRef]
43. Talari, S.; Yazdaninejad, M.; Haghifam, M.R. Stochastic-based scheduling of the microgrid operation including wind turbines, photovoltaic cells, energy storages and responsive loads. *IET Gener. Transm. Distrib.* **2015**, *9*, 1498–1509. [CrossRef]
44. Pourmousavi, S.A.; Nehrir, M.H.; Sharma, R.K. Multi-Timescale Power Management for Islanded Microgrids Including Storage and Demand Response. *IEEE Trans. Smart Grid* **2015**, *6*, 1185–1195. [CrossRef]
45. Rosenthal, R.E. *GAMS—A User's Guide*; ALS-NSCORT: West Lafayette, IN, USA, 2004.
46. Soroudi, A.; Amraee, T. Decision making under uncertainty in energy systems: State of the art. *Renew. Sustain. Energy Rev.* **2013**, *28*, 376–384. [CrossRef]
47. Tascikaraoglu, A.; Uzunoglu, M. A review of combined approaches for prediction of short-term wind speed and power. *Renew. Sustain. Energy Rev.* **2014**, *34*, 243–254. [CrossRef]
48. Contreras, J.; Espinola, R.; Nogales, F.J.; Conejo, A.J. ARIMA models to predict next-day electricity prices. *IEEE Trans. Power Syst.* **2003**, *18*, 1014–1020. [CrossRef]
49. Garcia, R.C.; Contreras, J.; Akkeren, M.; Garcia, J.B.C. A GARCH forecasting model to predict day-ahead electricity prices. *IEEE Trans. Power Syst.* **2005**, *20*, 867–874. [CrossRef]

50. Teo, K.K.; Wang, L.; Lin, Z. Wavelet Packet Multi-layer Perceptron for Chaotic Time Series Prediction: Effects of Weight Initialization. In Proceedings of the Computational Science—ICCS 2001: International Conference, San Francisco, CA, USA, 28–30 May 2001; pp. 310–317.
51. Al-Fattah, S.M. Artificial Neural Network Models for Forecasting Global Oil Market Volatility. *SSRN Electron. J.* **2013**, *112*. [CrossRef]
52. Amjady, N. Day-ahead price forecasting of electricity markets by a new fuzzy neural network. *IEEE Trans. Power Syst.* **2006**, *21*, 887–896. [CrossRef]
53. Che, J.; Wang, J. Short-term electricity prices forecasting based on support vector regression and Auto-regressive integrated moving average modeling. *Energy Convers. Manag.* **2010**, *51*, 1911–1917. [CrossRef]
54. Aien, M.; Hajebrahimi, A.; Fotuhi-Firuzabad, M. A comprehensive review on uncertainty modeling techniques in power system studies. *Renew. Sustain. Energy Rev.* **2016**, *57*, 1077–1089. [CrossRef]
55. Aien, M.; Rashidinejad, M.; Fotuhi-Firuzabad, M. On possibilistic and probabilistic uncertainty assessment of power flow problem: A review and a new approach. *Renew. Sustain. Energy Rev.* **2014**, *37*, 883–895. [CrossRef]
56. Mirakyan, A.; De Guio, R. Modelling and uncertainties in integrated energy planning. *Renew. Sustain. Energy Rev.* **2015**, *46*, 62–69. [CrossRef]
57. Yan, J.; Liu, Y.; Han, S.; Wang, Y.; Feng, S. Reviews on uncertainty analysis of wind power forecasting. *Renew. Sustain. Energy Rev.* **2015**, *52*, 1322–1330. [CrossRef]
58. Marino, C.; Quddus, M.A.; Marufuzzaman, M.; Cowan, M.; Bednar, A.E. A chance-constrained two-stage stochastic programming model for reliable microgrid operations under power demand uncertainty. *Sustain. Energy Grids Netw.* **2018**, *13*, 66–77. [CrossRef]
59. Dantzig, G.B. Linear programming under uncertainty. In *Stochastic Programming*; Springer: New York, NY, USA, 2011; pp. 1–11.
60. Xie, S.; Hu, Z.; Zhou, D.; Li, Y.; Kong, S.; Lin, W.; Zheng, Y. Multi-objective active distribution networks expansion planning by scenario-based stochastic programming considering uncertain and random weight of network. *Appl. Energy* **2018**, *219*, 207–225. [CrossRef]
61. Wang, Z.; Shen, C.; Liu, F. A conditional model of wind power forecast errors and its application in scenario generation. *Appl. Energy* **2018**, *212*, 771–785. [CrossRef]
62. Shields, M.D. Adaptive Monte Carlo analysis for strongly nonlinear stochastic systems. *Reliab. Eng. Syst. Saf.* **2018**, *175*, 207–224. [CrossRef]
63. Dufo-López, R.; Pérez-Cebollada, E.; Bernal-Agustín, J.L.; Martínez-Ruiz, I. Optimisation of energy supply at off-grid healthcare facilities using Monte Carlo simulation. *Energy Convers. Manag.* **2016**, *113*, 321–330. [CrossRef]
64. Caralis, G.; Diakoulaki, D.; Yang, P.; Gao, Z.; Zervos, A.; Rados, K. Profitability of wind energy investments in China using a Monte Carlo approach for the treatment of uncertainties. *Renew. Sustain. Energy Rev.* **2014**, *40*, 224–236. [CrossRef]
65. Bordbari, M.J.; Seifi, A.R.; Rastegar, M. Probabilistic energy consumption analysis in buildings using point estimate method. *Energy* **2018**, *142*, 716–722. [CrossRef]
66. Hong, H.P. An efficient point estimate method for probabilistic analysis. *Reliab. Eng. Syst. Saf.* **1998**, *59*, 261–267. [CrossRef]
67. Peik-Herfeh, M.; Seifi, H.; Sheikh-El-Eslami, M.K. Decision making of a virtual power plant under uncertainties for bidding in a day-ahead market using point estimate method. *Int. J. Electr. Power Energy Syst.* **2013**, *44*, 88–98. [CrossRef]
68. Marandi, S.M.; Anvar, M.; Bahrami, M. Uncertainty analysis of safety factor of embankment built on stone column improved soft soil using fuzzy logic α-cut technique. *Comput. Geotech.* **2016**, *75*, 135–144. [CrossRef]
69. Suganthi, L.; Iniyan, S.; Samuel, A.A. Applications of fuzzy logic in renewable energy systems—A review. *Renew. Sustain. Energy Rev.* **2015**, *48*, 585–607. [CrossRef]
70. Soroudi, A.; Ehsan, M. A possibilistic–probabilistic tool for evaluating the impact of stochastic renewable and controllable power generation on energy losses in distribution networks—A case study. *Renew. Sustain. Energy Rev.* **2011**, *15*, 794–800. [CrossRef]
71. Zhang, H.; Yue, D.; Xie, X. Robust Optimization for Dynamic Economic Dispatch Under Wind Power Uncertainty with Different Levels of Uncertainty Budget. *IEEE Access* **2016**, *4*, 7633–7644. [CrossRef]

72. Kuznetsova, E.; Ruiz, C.; Li, Y.-F.; Zio, E. Analysis of robust optimization for decentralized microgrid energy management under uncertainty. *Int. J. Electr. Power Energy Syst.* **2015**, *64*, 815–832. [CrossRef]
73. Zare, K.; Conejo, A.J.; Carrión, M.; Moghaddam, M.P. Multi-market energy procurement for a large consumer using a risk-aversion procedure. *Electr. Power Syst. Res.* **2010**, *80*, 63–70. [CrossRef]
74. Alipour, M.; Zare, K.; Mohammadi-Ivatloo, B. Optimal risk-constrained participation of industrial cogeneration systems in the day-ahead energy markets. *Renew. Sustain. Energy Rev.* **2016**, *60*, 421–432. [CrossRef]
75. Nojavan, S.; Ghesmati, H.; Zare, K. Robust optimal offering strategy of large consumer using IGDT considering demand response programs. *Electr. Power Syst. Res.* **2016**, *130*, 46–58. [CrossRef]
76. Fu, Q.; Nasiri, A.; Bhavaraju, V.; Solanki, A.; Abdallah, T.; Yu, D.C. Transition Management of Microgrids With High Penetration of Renewable Energy. *IEEE Trans. Smart Grid* **2014**, *5*, 539–549. [CrossRef]
77. Liu, Y.; Yu, S.; Zhu, Y.; Wang, D.; Liu, J. Modeling, planning, application and management of energy systems for isolated areas: A review. *Renew. Sustain. Energy Rev.* **2018**, *82*, 460–470. [CrossRef]
78. Mazzola, S.; Astolfi, M.; Macchi, E. A detailed model for the optimal management of a multigood microgrid. *Appl. Energy* **2015**, *154*, 862–873. [CrossRef]
79. Chen, C.; Duan, S.; Cai, T.; Liu, B.; Hu, G. Smart energy management system for optimal microgrid economic operation. *IET Renew. Power Gener.* **2011**, *5*, 258–267. [CrossRef]
80. Patra, S.B.; Mitra, J.; Ranade, S.J. Microgrid architecture: A reliability constrained approach. In Proceedings of the IEEE Power Engineering Society General Meeting, San Francisco, CA, USA, 16 June 2005; pp. 2372–2377.
81. Mitra, J.; Patra, S.B.; Ranade, S.J. A dynamic programming based method for developing optimal microgrid architectures. In Proceedings of the 15th Power Systems Computational Conference, Liège, Belgium, 22–26 August 2005; pp. 22–26.
82. Nosratabadi, S.M.; Hooshmand, R.-A.; Gholipour, E. A comprehensive review on microgrid and virtual power plant concepts employed for distributed energy resources scheduling in power systems. *Renew. Sustain. Energy Rev.* **2017**, *67*, 341–363. [CrossRef]
83. Marzband, M.; Azarinejadian, F.; Savaghebi, M.; Guerrero, J.M. An Optimal Energy Management System for Islanded Microgrids Based on Multiperiod Artificial Bee Colony Combined with Markov Chain. *IEEE Syst. J.* **2015**, *11*, 1712–1722. [CrossRef]
84. Huynh, T.D.; Jennings, N.R.; Shadbolt, N.R. An integrated trust and reputation model for open multi-agent systems. *Auton. Agents Multi-Agent Syst.* **2006**, *13*, 119–154. [CrossRef]
85. Vytelingum, P.; Voice, T.D.; Ramchurn, S.D.; Rogers, A.; Jennings, N.R. Agent-based micro-storage management for the smart grid. In Proceedings of the 9th International Conference on Autonomous Agents and Multiagent Systems, Toronto, ON, Canada, 10–14 May 2010; pp. 39–46.
86. Dehghanpour, K.; Colson, C.; Nehrir, H. A survey on smart agent-based microgrids for resilient/self-healing grids. *Energies* **2017**, *10*, 620. [CrossRef]
87. Lezama, F.; Palominos, J.; Rodríguez-González, A.Y.; Farinelli, A.; de Cote, E.M. Optimal Scheduling of On/Off Cycles: A Decentralized IoT-Microgrid Approach. In *Applications for Future Internet*; Springer: Berlin, Germany, 2017; pp. 79–90.
88. Ferdinando, F.; Yeoh, W.; Pontelli, E. Multi-variable agents decomposition for DCOPs. In Proceedings of the Thirtieth AAAI Conference on Artificial Intelligence, Phoenix, AZ, USA, 12–17 February 2016.
89. Gurobi Optimizer Inc. *Gurobi Optimizer Reference Manual*; Gurobi Optimization Inc.: Houston, TX, USA, 2014.
90. Unamuno, E.; Barrena, J.A. Hybrid ac/dc microgrids—Part I: Review and classification of topologies. *Renew. Sustain. Energy Rev.* **2015**, *52*, 1251–1259. [CrossRef]
91. Barnes, M.; Kondoh, J.; Asano, H.; Oyarzabal, J.; Ventakaramanan, G.; Lasseter, R.; Hatziargyriou, N.; Green, T. Real-World MicroGrids-An Overview. In Proceedings of the IEEE International Conference on System of Systems Engineering, San Antonio, TX, USA, 16–18 April 2007; pp. 1–8.
92. Hossain, E.; Kabalci, E.; Bayindir, R.; Perez, R. Microgrid testbeds around the world: State of art. *Energy Convers. Manag.* **2014**, *86*, 132–153. [CrossRef]
93. Jiang, Q.; Xue, M.; Geng, G. Energy Management of Microgrid in Grid-Connected and Stand-Alone Modes. *IEEE Trans. Power Syst.* **2013**, *28*, 3380–3389. [CrossRef]
94. Tsikalakis, A.G.; Hatziargyriou, N.D. Centralized Control for Optimizing Microgrids Operation. *IEEE Trans. Energy Convers.* **2008**, *23*, 241–248. [CrossRef]

95. Kargarian, A.; Falahati, B.; Fu, Y.; Baradar, M. Multiobjective optimal power flow algorithm to enhance multi-microgrids performance incorporating IPFC. In Proceedings of the IEEE Power and Energy Society General Meeting, San Diego, CA, USA, 22–26 July 2012; pp. 1–6.
96. Nunna, H.S.V.; Doolla, S. Demand Response in Smart Distribution System with Multiple Microgrids. *IEEE Trans. Smart Grid* **2012**, *3*, 1641–1649. [CrossRef]
97. Wang, Z.; Chen, B.; Wang, J.; Begovic, M.M.; Chen, C. Coordinated Energy Management of Networked Microgrids in Distribution Systems. *IEEE Trans. Smart Grid* **2015**, *6*, 45–53. [CrossRef]

Article

An Augmented Prony Method for Power System Oscillation Analysis Using Synchrophasor Data

Moossa Khodadadi Arpanahi [1], Meysam Kordi [2], Roozbeh Torkzadeh [3], Hassan Haes Alhelou [1,4] and Pierluigi Siano [5,*]

1 Department of Electrical and Computer Engineering, Isfahan University of Technology (IUT), Isfahan 84156-83111, Iran; moossa.khodadadi@ec.iut.ac.ir (M.K.A.); h.haesalhelou@gmail.com or h.haesalhelou@ec.iut.ac.ir (H.H.A.)
2 Intelliwide Company, Ergonomic and Capable Products for Power Systems, Isfahan 8173986561, Iran; kordi@intelliwide.com
3 Institute of Science and Technology, Universidad Loyola Andalucía, Campus de Palmas Altas, No. 1, Energy Solar St., 41014 Seville, Spain; roozbeh.torkzadeh@gmail.com
4 Department of Electrical Power Engineering, Faculty of Mechanical and Electrical Engineering, Tishreen University, Lattakia 2230, Syria
5 Department of Management & Innovation Systems, University of Salerno, 84084 Salerno, Italy
* Correspondence: psiano@unisa.it; Tel.: +39-089-96-4294

Received: 8 March 2019; Accepted: 29 March 2019; Published: 2 April 2019

Abstract: Intrinsic mode functions (IMFs) provide an intuitive representation of the oscillatory modes and are mainly calculated using Hilbert–Huang transform (HHT) methods. Those methods, however, suffer from the end effects, mode-mixing and Gibbs phenomena since they use an iterative procedure. This paper proposes an augmented Prony method for power system oscillation analysis using synchrophasor data obtained from a wide-area measurement system (WAMS). In the proposed method, in addition to the estimation of the modal information, IMFs are extracted using a new explicit mathematical formulation. Further, an indicator based on an energy and phase relationship of IMFs is proposed, which allows system operators to recognize the most effective generators/actuators on specific modes. The method is employed as an online oscillation-monitoring framework providing inputs for the so-called wide-area damping control (WADC) module. The efficacy of the proposed method is validated using three test cases, in which the IMFs calculation is simpler and more accurate if compared with other methods.

Keywords: power system oscillations; stability; Prony method; intrinsic mode functions (IMFs); wide-area measurement system (WAMS); phasor measurement units (PMUs)

---

1. Introduction

Power system oscillation analysis has received considerable attention over the past decades. Features such as frequency and damping of the power system oscillations provide critical information for system operators [1]. Analysis of such oscillations are conducted using: (i) model-based and (ii) measurement-based approaches. The former approach is based on the linearized differential equation of the system around a specific operating point [2,3], as accurate modeling of power systems for various operating conditions is complicated and these approaches may not be useful for practical applications. On the other hand, with developing phasor measurement units (PMUs) and wide-area measurement system (WAMS), the latter approach became more popular. Recently, several studies have attempted to extract oscillation parameters using synchrophasor data. Some of the well-known methods used for oscillation detection using measurement-based approaches include: Kalman Filter (KF) [4] estimation of signal parameters by rotational invariance technique

(ESPRIT) [5], Hilbert–Huang transform (HHT) [6–9], Yule–Walker algorithm [10], wavelet transform (WT) [11], recursive adaptive stochastic subspace identification (RASSI) [12], maximum likelihood method [13], and Prony analysis method [14–16]. Among them, the Prony method is one of the most commonly used measurement-based methods. The Prony method can directly estimate modal components of a measured signal. Recently, extensions of the Prony method have been developed that can calculate oscillatory modes of multiple signals in a central [14] or distributed [15] manner. Furthermore, a forward and backward extended Prony (FBEP) method is proposed in order [16] to reduce the sensitivity of the Prony method to noise.

A practical online oscillation monitoring framework should be able to determine the power system oscillations fast and accurately. Further, it should be capable of identifying contributions of different system buses in each oscillatory mode. Among the above methods, only the HHT method can determine the contribution of each state variable in various oscillatory modes. In general, each state variable of the system can be expressed as the sum of exponential sinusoids. According to the HHT method, each of the exponential sinusoids is called intrinsic mode functions (IMF). Using IMFs, participation of each system bus in an oscillatory mode can be determined by means of an indicator called node contribution factor (NCF). In the HHT method, IMFs can be extracted from the synchrophasor data using empirical mode decomposition (EMD), which is an iterative procedure. However, EMD process has some inherent shortcomings such as the end effects (EEs), mode-mixing and Gibbs phenomena. Some of those drawbacks have been addressed in improved versions of the HHT method in [6–9].

In this paper, we propose an augmented Prony based framework for real-time power system oscillation monitoring and analysis using synchrophasor data. The proposed framework estimates modal features (i.e., its damping and frequency) of a measured signal, while it is capable of determining contributions of different system buses in oscillatory modes by calculating IMFs and NCFs. Note that in the proposed framework, IMFs and thus NCFs are extracted using an explicit mathematical formulation based on an augmented Prony method. To the best of the authors' knowledge, the Prony method has never been used before for the calculation of IMFs and thus NCFs. Unlike the HHT method that relies on heuristic approaches such as EMD for the calculation of IMFs, in the proposed approach they are calculated using an explicit mathematical formulation, which is one of the novelties of the proposed method. The calculation of IMFs in the proposed framework is more accurate, while it requires a lower computational burden if compared to the existing HHT methodologies. The proposed online oscillation-monitoring framework can be used as an input for the so-called wide-area damping control (WADC) module, which is responsible for controlling the system oscillations. Note that obtaining NCFs associated with oscillatory modes in the system can assist power system operators to take better remedial actions enhancing system stability and security. The effectiveness of the proposed framework is validated using the Kundur's two-area test system and a practical power system. The main innovative contributions in this paper are:

- A novel augmented Prony-based oscillation analysis method is proposed for smart systems with WAMS.
- A novel explicit mathematical formulation is proposed for online obtaining IMFs and NCFs.
- A new IMF-based method for determining the effective order of the system is also suggested.
- The proposed method can eliminate the negative effects of EE, Gibbs, and mode-mixing which are the main challenges in previous methods such as HHT.
- The proposed NCF can help system operators to recognize the most effective generators/actuators on a specific mode in the system.

The rest of the paper is organized as follows: the motivations behind the work and the formulations IMFs and NCFs calculation in the proposed framework are presented in Section 2. In Section 3, performance of the proposed framework is analyzed using the Kundur's two-area power system and a practical power system. Finally, Section 4 concludes the paper.

2. Proposed Framework for WAMS-Based Oscillation Detection

2.1. Research Motivation

Developing a real-time power system oscillation monitoring framework is an important part of small signal stability studies using WAMS technologies. An overview of Smart Grid components and technologies is shown in Figure 1. Observe that the smart grid consists of different state-of-the-art technologies enabling effective network management schemes in the presence of renewable energy systems, flexible loads, and energy storage systems. This new concept ultimately offers the efficient and reliable operation of future power systems via a two-way exchange of data and energy among different sectors in the electric power supply chainmn including generation, transmission, distribution, and consumption. Implementation of PMUs and WAMS to achieve a fully observable grid was defined as the first phase of a Smart Grid project [17,18].

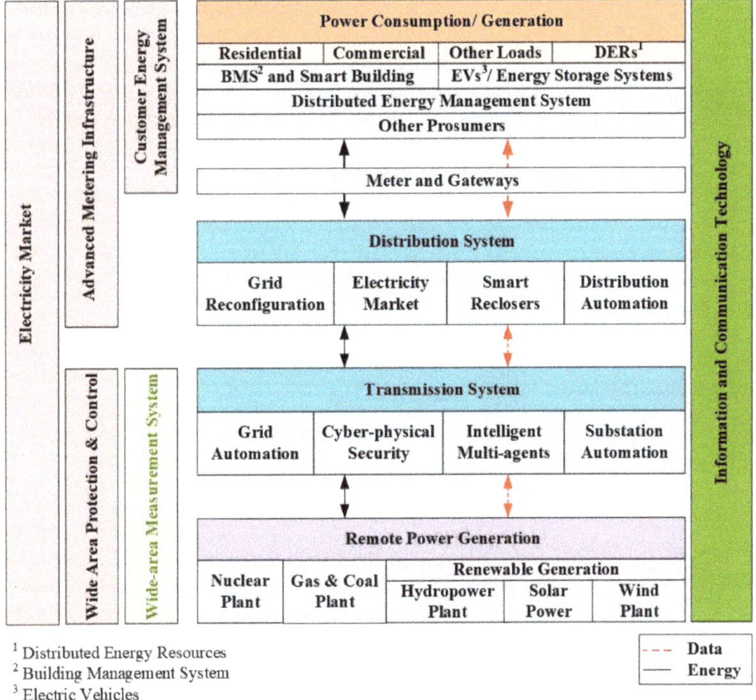

Figure 1. General outline of a Smart Grid.

As shown in Figure 2, WAMS architecture includes five layers. The component layer consists of physical devices. It is responsible for providing data, functions, and communication services for the upper layers. The communication layer applies several methods and protocols to obtain data among various elements in the network. The information layer includes the data model and the communication rules that are used to exchange information. The function layer contains the logical functions or various WAMS applications, independent of the physical architecture; however, it is reliant on the business models and the organization's policies. Finally, the business layer defines the business models and the policies [17,18]

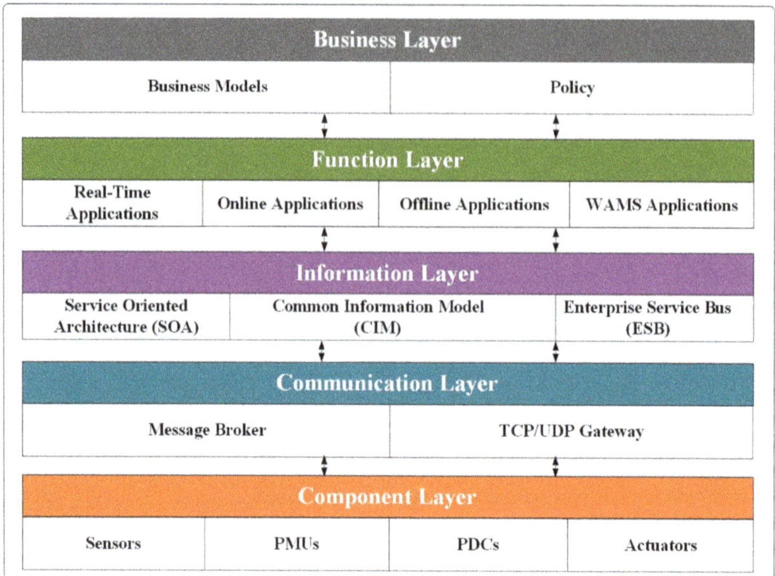

Figure 2. Wide-area measurement system (WAMS) architecture.

One of the main online applications associated with the proposed WAMS function layer is power system oscillation detection using synchrophasor data, which is the main scope of this paper. The oscillation detection algorithm is detailed in Sections 2.2–2.5.

2.2. Motivation behind Methodology Selection

In many Smart Grid projects, some well-known methods used for power system oscillation detection were analyzed and compared. This has been carried out to identify the best method in terms of: (1) accuracy in the estimation of the oscillatory mode features, (2) simplicity in the implementation, (3) ability to provide visual intuitiveness, (4) skill to determine the most associated state variable to each oscillatory mode, (5) robustness against the noise, and (6) self-verification ability. A survey of the literature showed that the extended Prony methods can be a reasonable choice for the proposed framework. Note that the basic Prony method has some weakness such as its sensitivity to noise and selection of the proper order of the system [14–16]. An overestimated order may lead to the generation of extraneous modes by the Prony method. These drawbacks were addressed in some extensions of the Prony method. For example, the study in [15] proposed a simple criteria for selecting the order of the system. The study in [16] suggested that the Prony method can be performed in forward and backward manners, where the final modes are selected from the collection of those extracted in such a process. This process enabled the Prony method to identify extraneous modes, while allowed elimination of the noise effects. Prony is inherently a self-verifiable method because it estimates the modes of a given signal and then reconstructs the main signal from the estimated mode. Comparing the main and the reconstructed signal is a convenient way to verify its performance.

The above features make the Prony method an effective one for practical applications; however, it does not provide visual intuitiveness on contribution various system modes in each oscillatory modes. This is important because power system operators are interested in visual information instead of numerical data. In this regard, the HHT method has advantages over the Prony method, in which the so-called IMFs plots represent the behavior of each mode intuitively, and therefore provide situational awareness on oscillations of the power system. Moreover, the HHT method can determine the contribution of a state variable in a mode based on energy and phase relationship of IMFs. Nevertheless, as stated in the introduction, the HHT method suffers from the EEs, mode-mixing and Gibbs phenomena [6,9]. In this paper, an improved Prony method is proposed that is able to calculate IMFs and thus NCFs for a given signal.

Note that HHT has a much wider application than the Prony method and the authors don't mean the proposed method is generally better than HHT, especially in non-linear situations. We compare our method with HHT because it is the only method that can calculate IMFs and NCFs. To achieve the IMFs and NCFs, HHT should overcome some challenges such as end-effects, mode mixing, and Gibbs phenomena. These challenges are solved in different literature. For example, there are several ways to overcome end-effect issue [17]. In this paper, a direct method is proposed to obtain the IMFs and NCFs, in which the HHT challenges are avoided. Indeed, we are facing the alternative to obtain IMFs and NCFs as depicted in the following figure:

1. HHT method that obtains IMFs and NCFs carefully, after solving the end-effects, mode mixing, and Gibbs phenomena. It is based on a heuristic, iterative method. Although it can handle non-linear situations [18], however, it is not the scope of this paper. Since we deal with the small signal stability assessment in which the linearization is an acceptable and reasonable assumption.
2. Our Proposed method, which obtains IMFs and NCFs carefully, directly and without need to solve the end-effects, mode mixing, and Gibbs phenomena. It is based on an explicit mathematical formulation. It can be used for applications such as oscillation analysis, which is scope of this paper.

Therefore, both methods (HHT and proposed method) can handle the IMFs and NCFs with different approaches and can be used based on the situations. Figure 3, briefly shows the above discussions.

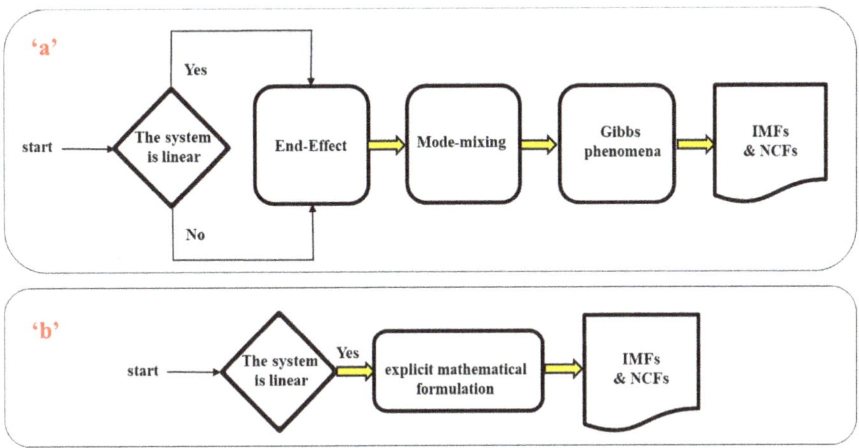

Figure 3. General outline of (**a**) Hilbert–Huang transform (HHT) method; (**b**) proposed method.

2.3. Oscillation Analysis Framework

Assuming a power system of order n, the zero-input time response is as follows [14]:

$$\underline{y}(t) = \sum_{i=1}^{n} \gamma_i e^{\lambda_i t} \tag{1}$$

where λ_i are called eigenvalues or modes of the system. The Prony method represents a given signal as a sum of exponential functions and then estimates the unknown parameters, i.e., γ_i and λ_i so that the real and the estimated output have a minimum difference. The estimated signal in the discrete time domain, and can be expressed as follows [14]:

$$\underline{\hat{y}}(k) = \sum_{i=1}^{n} \beta_i {z_i}^k \quad k = 1, 2 \ldots, N \tag{2}$$

where, z_i are the eigenvalues of the system in the discrete time domain or z-domain, β_i is the residue corresponding to z_i and N is the number of the samples. Equation (2) can be described in a matrix form as [14]:

$$\begin{bmatrix} {z_1}^{1_s} & {z_2}^{1_s} & \cdots & {z_n}^{1} \\ {z_1}^{2_s} & {z_2}^{2_s} & \cdots & {z_n}^{2_s} \\ \vdots_1^{N_s} & \vdots_2^{N_s} & \vdots & \vdots_N^{N_s} \\ {z_1}^N & {z_2}^N & \cdots & {z_n}^N \end{bmatrix} \begin{bmatrix} \beta_1 \\ \beta_2 \\ \vdots \\ \beta_n \end{bmatrix} = \begin{bmatrix} y(1) \\ y(2) \\ \vdots \\ y(N) \end{bmatrix} \tag{3}$$

According to the Prony method, for a discrete time system with an output as (2), the so called characteristic equation is as follows [14]:

$$z^n - a_1 z^{n-1} - a_2 z^{n-2} - \cdots - a_{n-1} z - a_n = 0 \tag{4}$$

The coefficients of the above equation are obtained by solving the predictive equation as follows [14]:

$$\begin{bmatrix} y(n) & \cdots & y(1) \\ y(n+1) & \cdots & y(2) \\ \vdots & \vdots & \vdots \\ y(N-1) & \cdots & y(N-n) \end{bmatrix} \begin{bmatrix} a_1 \\ a_2 \\ \vdots \\ a_N \end{bmatrix} = \begin{bmatrix} y(n+1) \\ y(n+2) \\ \vdots \\ y(N) \end{bmatrix} \tag{5}$$

which can be given in a matrix form as:

$$D\underline{a} = \underline{d} \tag{6}$$

where, D is the Prony Matrix. For WAMS-monitored power systems, both D and $\underline{d}$ contain synchrophasor data (e.g., voltages, currents, or powers). Based on the above formulations, the algorithm of the Prony method can be summarized as follows [14–16]:

1. Constitute D and $\underline{d}$ using PMU data according to (5).
2. Solve a linear least square error (LSE) problem and obtain vector $\underline{a}$.
3. Solve the polynomial characteristic Equation (4) to achieve the z-domain eigenvalues.
4. Solve another LSE problem (3), using z values in Step 3.
5. Reconstruct the main signal using (2).

Based on the above algorithm, in Step 3, the z-domain eigenvalues are calculated. The eigenvalues of the main signal can be obtained as follows [14]:

$$z_i = e^{\lambda_i T} \Rightarrow \lambda_i = \frac{log(z_i)}{T} \quad , i = 1, 2 \ldots, n \tag{7}$$

where, T is the sampling rate of the main signal. In many cases, instead of complex values of eigenvalues, two well-known parameters of the modes, i.e., damping ratio ξ and damping frequency f_d are represented, which are related to eigenvalues as follows [1]:

$$\lambda_i = \sigma_i \pm j\omega_{d_i} \ , \xi_i = \frac{\sigma_i}{|\lambda_i|} \ , \ f_{d_i} = \frac{\omega_{d_i}}{2\pi} \quad , i = 1, 2 \ldots, n \tag{8}$$

2.4. Calculation of IMFs and NCFs Based on Augmented Prony Method

As previously stated, an online oscillation monitoring framework should be able to determine the contribution of various system buses in each oscillatory mode. Here, as one of the novel contributions of this work, it is demonstrated that the Prony method can be used for such a purpose. Initially, the matrix U is defined as follows:

$$U = \begin{bmatrix} z_1{}^{1_s} & z_2{}^{1_s} & \cdots & z_n{}^{1} \\ z_1{}^{2_s} & z_2{}^{2_s} & \cdots & z_n{}^{2_s} \\ \vdots_1{}^{N_s} & \vdots_2{}^{N_s} & \vdots & \vdots_N{}^{N_s} \\ z_1{}^{N} & z_2{}^{N} & \cdots & z_n{}^{N} \end{bmatrix} \tag{9}$$

where, u_i is defined as column i of matrix U, thus, for a complex eigenvalue, there is another eigenvalue which is the conjugate of z_i. Without the loss of generality, it is assumed that z_{i+1} and β_{i+1} are the conjugates of z_i and β_i, respectively ($z_{i+1} = z_i{}^*$; $\beta_{i+1} = \beta_i{}^*$), thus:

$$\beta_i\, u_i + \beta_{i+1}\, u_{i+1} = \beta_i\, u_i + \beta_i{}^* u_i{}^* \quad ,i = 1,2\ldots,n \tag{10}$$

Substituting u_i from (9) into (10) yields:

$$\beta_i\, u_i + \beta_i{}^* u_i{}^* = \beta_i \begin{bmatrix} z_i{}^1 \\ z_i{}^2 \\ \vdots \\ z_i{}^k \\ \vdots \\ z_i{}^N \end{bmatrix} + \beta_i{}^* \begin{bmatrix} (z_i{}^*)^1 \\ (z_i{}^*)^2 \\ \vdots \\ (z_i{}^*)^k \\ \vdots \\ (z_i{}^*)^N \end{bmatrix} \quad ,i = 1,2\ldots,n \tag{11}$$

where, β_i and z_i are defined as follows:

$$z_i = e^{\lambda_i T} = e^{(\sigma_i + j\omega_{di})T}\, , \beta_i = \frac{A_i}{2} e^{j\varphi_i} \quad ,i = 1,2\ldots,n \tag{12}$$

Thus, (11) can be rewritten as:

$$\beta_i\, \mu_i + \beta_i{}^* \mu_i{}^* = \begin{bmatrix} A_i e^{\delta_i T} cos[\omega_{d_i} T + \varphi_i] \\ A_i e^{\delta_i (2T)} cos[\omega_{d_i} (2T) + \varphi_i] \\ \vdots \\ A_i e^{\delta_i (kT)} cos[\omega_{d_i} (kT) + \varphi_i] \\ \vdots \\ A_i e^{\delta_i (NT)} cos[\omega_{d_i} (NT) + \varphi_i] \end{bmatrix}, \; i = 1,2\ldots,n \tag{13}$$

According to the definition of IMF (each of the exponential sinusoids contained in the main signal), the kth row of the vector in (13) shows the value of IMF of mode i at the time $t = kT$. Thus, IMFs can be calculated as in (14):

$$IMF_i(t = kT) = \beta_i\, u_i(k) + \beta_i{}^* u_i{}^*(k) \quad ,i = 1,2\ldots,n \tag{14}$$

The above expression is valid for complex modes. For any real mode, its corresponding column in matrix U results in values of IMF_i. Therefore, a generalized form for the calculation of IMFs can be expressed as:

$$IMF_i(t = kT) = \beta_i u_i(k) + c_i \beta_i{}^* u_i{}^*(k), \quad i = 1,2\ldots,n \tag{15}$$

where, c_i is a binary variable that shows either mode i is complex ($c_i = 1$) or not ($c_i = 0$). Note that it is assumed in (15) that columns of the matrix U are sorted based on their real part.

Having intrinsic function of each mode, it is possible to determine the most associated state variable corresponding to each oscillatory mode, using energy and phase relationship of IMFs. In other words, the generator or actuator that is the most associated ones with a specific mode can be identified. As mentioned in the Introduction, the participation of each system bus in an oscillatory mode can be determined using node contribution factor (NCF). This index is defined as follows:

$$NCF_{s,i} = E_{s,i} cos(\varphi_{s,i}) \tag{16}$$

where, $NCF_{s.i}$ represents the contribution of signal s in mode i. $E_{s,i}$ is defined as the energy of IMF_{i} (related to mode i) obtained by signal s, which is defined as follows:

$$E_{s,i} = \sum_{k=1}^{N} IMF_{s,i}{}^2(t = kT) \tag{17}$$

In (16), $\varphi_{s,i}$ is the average relative phase of $IMF_{s,i}$ and the first IMF ($IMF_{s,1}$ that is associated with the dominant mode) in different times:

$$\varphi_{s,i} = \left(\frac{1}{N}\right) \cdot \sum_{t=1}^{N} \theta_{s,i}(t) - \theta_{s,1}(t) \tag{18}$$

In order to calculate $\varphi_{s,i}$ a phase angle is assigned to each data point of $IMF_{s,i}$. For all IMFs extracted from a measured signal of the network, the initial phase angle of the first IMF is defined as the reference phase ($\theta_{s,1}(t) = 0$). For a specific IMF, the phase angle of maximum, minimum, positive zero-crossing, and the negative zero-crossing points are set as $\pi/2$, $3\pi/2$, 0 (or 2π) and π, respectively. According to the proposed formulation in (15)–(18), it is possible to directly calculate IMFs of the various signals of the system and therefore obtain their contributions in the different oscillatory modes. It is important to emphasize that so far, the Prony method has not been augmented for calculation of IMFs and thus NCFs. Our proposed formulation in this subsection bypasses the problems of nonlinear and the iterative method of HHT method for IMF calculation, such as the mode mixing of EMD, end effect and Gibbs phenomena [6–9].

Note that the understudy signal may contain noise that can be easily eliminated before performing the proposed method. It is worth mentioning that the mode mixing phenomena is not related to the noise. In the HHT method the IMFs are obtained using an iterative procedure with two loops, the outer loop repeats with the number of modes and for each mode an iterative procedure called sifting is performed to achieve the IMF associated with that mode. When the frequency of two modes of the system is close to each other, mode mixing may have occurred. In fact it is related to frequency heterodyne. Therefore, noise that is a high frequency signal may not mix with power system modes, which are low-frequency modes. In our proposed method, all sifting or decomposition procedures are not performed; therefore it is not prone to mode mixing. In the proposed method, each mode has a unique IMF. Thus, with an explicit mathematical formulation, each mode and its IMF are specifically extracted.

It is worth mentioning that power system oscillation detection and analysis is one of the topics of small signal stability assessment. In small signal stability assessment and analysis, it is generally accepted that the variation of the state variables of the system around the operating point are small, thus the differential equations of the system can reasonably be linearized at the operating point. The zero-input time response of such equations is seen in Equation (1). Therefore, despite the HHT method can handle signals without assumption of structure in Equation (1), in small signal stability analysis, which is the scope of this paper, we deal with the exponential sinusoids signals called intrinsic mode functions (IMFs). It is worth mentioning that the method is performed on many different signals with different parameters (amplitude, damping, frequency, and phase angle), and in all cases, the performance of the method in estimation of the mode parameters and extraction of the IMFs, are approved. However, an example of such signal is presented in Section 3.1.

2.5. Approximation System Order

In practical situations, the number of modes incorporated in the signal is unknown. Several methods are proposed in the literature to approximate the effective order of the system [14–16]. In this paper, a new IMF-based index is introduced to determine the effective system order. The smallest index m that satisfies the inequality in (18) is the effective order of the system.

$$\frac{\sum_{i=1}^{m} IMF_{s,i}{}^{2}}{\sum_{i=1}^{n} IMF_{s,i}{}^{2}} \geq \tau \tag{19}$$

where, the threshold value τ is close to 1 (e.g., 0.999) and n is the initial value of the system order.

3. Simulation Results and Discussion

3.1. Test Case 1: Hypothetical Signal

In this test case, a hypothetical signal give in (20) with four pairs of complex conjugate eigenvalues is considered. Parameters of this signal are shown in Table 1. The main signal (i.e., $y(t)$) and its components (i.e., $y_1(t)$ to $y_4(t)$) are shown in Figure 4. It is assumed that signal $y(t)$ is measured by PMU and resampled at 10 Hz used as an input of our proposed method. The main advantage of this test case is that the real modes and their corresponding IMFs are predefined, thus it is suitable for the validation of the proposed method. The jth mode associated with $y(t)$ can be calculated according to (8). Also, $y_j(t)$ is by definition the jth IMF. The main signal is compared with the reconstructed signal by the Prony method in Figure 4. As shown in Figure 5, the reconstructed signal completely matches the real signal $y(t)$, as expected. As stated in Section 2.3, Figure 5 illustrates the self-verification property of the Prony method.

$$y(t) = \sum_{j=1}^{4} y_j(t) = \sum_{j=1}^{4} A_j e^{\sigma_j T} cos\left[2\pi f_{dj} t + \varphi_j\right] \tag{20}$$

Table 1. Parameters of the Hypothetical Signal of Test Case 1.

j	A_j	σ_j	f_{dj}	φ_j
1	1	−0.2	0.2	0
2	0.2	−1	1	π/2
3	−0.8	−0.7	0.8	π/6
4	2	−0.5	0.7	π/5

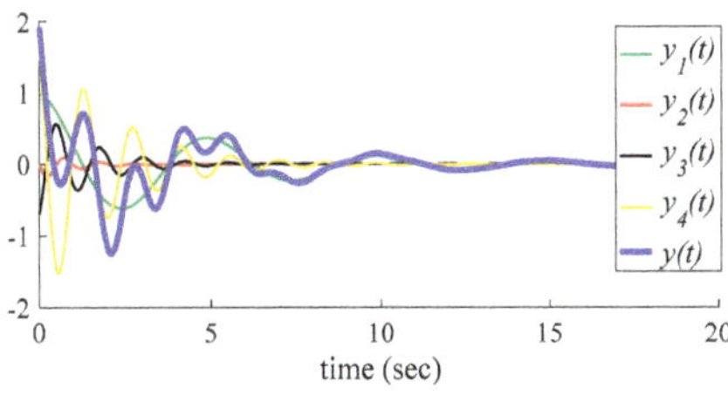

Figure 4. Hypothetical signal of Test Case 1 and its components.

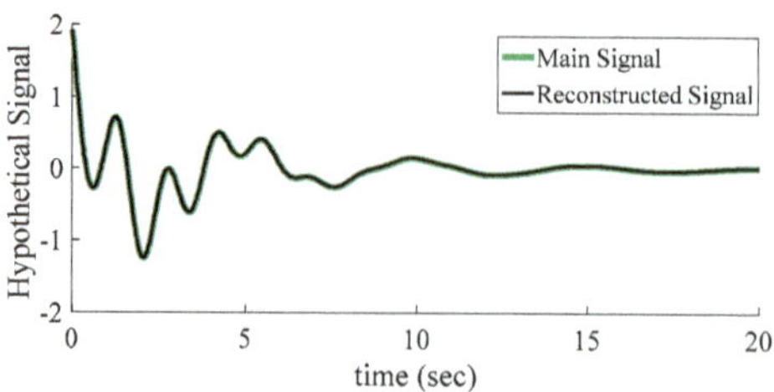

Figure 5. Comparison of real and reconstructed signals in Test Case 1.

The modes of the signal $y(t)$ obtained by the Prony method are compared with real modes in Table 2. Observe that the Prony method is capable to extract modes accurately. The IMFs of the hypothetical signal are shown in Figure 6. Observe that extracted IMFs from our proposed mathematical method in Section 2.4 are exactly the same as real IMFs $(y_j(t); j = 1, \ldots, 4)$. Thus, our mathematical formulation computes IMFs of a given signal with high accuracy, without a need for heuristic algorithms such as those employed in the HHT method. Also, as illustrated in Figure 6, the proposed method represents IMFs of a given signal, in order of increasing damping for them. In other words, IMF of a more dominant mode or equivalently less damped ones will be represented first in our methodology. Therefore, in the Test Case 1, IMF_1 to IMF_4 are corresponding to y_1, y_4, y_3, and y_2 in the hypothetical signal of (20), respectively, shown in Figure 4.

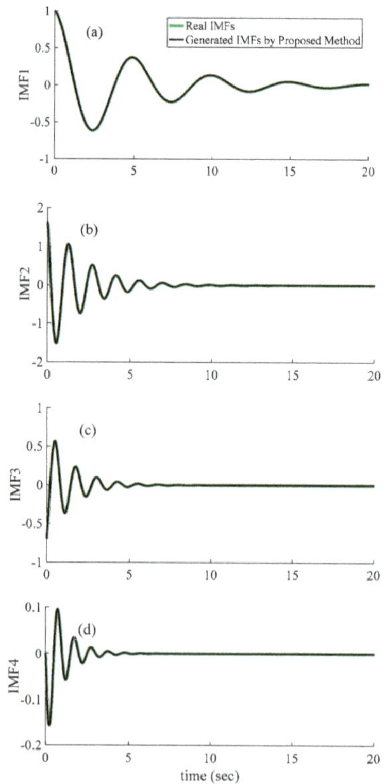

Figure 6. Comparison of real and estimated intrinsic mode functions (IMFs) of signal of Test Case 1. (**a–d**) IMF1 to IMF4, respectively.

Table 2. Calculated Modes of Test Case 1.

Real Modes		Estimated Modes	
$\xi(-)$	$f_d(Hz)$	$\xi(-)$	$f_d(Hz)$
0.157	0.2	0.157	0.2
0.113	0.7	0.113	0.7
0.138	0.8	0.138	0.8
0.157	1	0.157	1

3.2. Test Case 2: Kundur's Two-Area Power System

The second test case is carried out using Kundur's two-area test system [1], which is shown in Figure 7. These two areas are connected through weak transmission lines. The output active power of the generators denoted after the occurrence of a three-phase short-circuit fault is shown in Figure 8. The fault occurred at t = 1 s on Bus 8 and was cleared at t = 1.05 s. Assume that PG1 to PG4 are the output active power of the generators G1 to G4, respectively. Each of the four signals is analyzing using the proposed method in Sections 2.3 and 2.4 to extract modes of the system, their IMFs and consequently NCFs. The extracted modes based on each signal are depicted in Table 3. It is evident from the table that the extracted modes from the different signal (i.e., PG1 to PG4) result in the same modes. This is one way to validate the accuracy of the modes obtained by the Prony method. The main and reconstructed values of PG1 by the proposed method are compared in Figure 9, where again the accuracy of our framework is demonstrated. The extracted IMFs from generator G1 output power are demonstrated in Figure 10. Observe that IMF_1 is associated with an unstable mode, while the other three modes are under-damped. While analyzing numerical results of Table 3 is important, our framework provides visual information using IMF plots (using the proposed formulation in Section 2.4) such as the one shown in Figure 10, which provides situational awareness for oscillations during the fault.

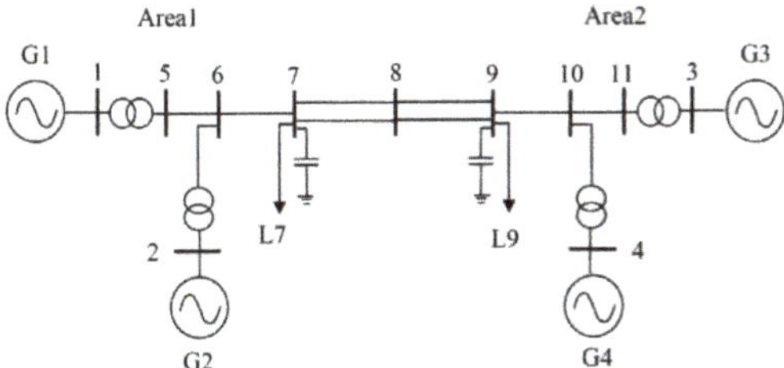

Figure 7. Kundur's two-area test system [1].

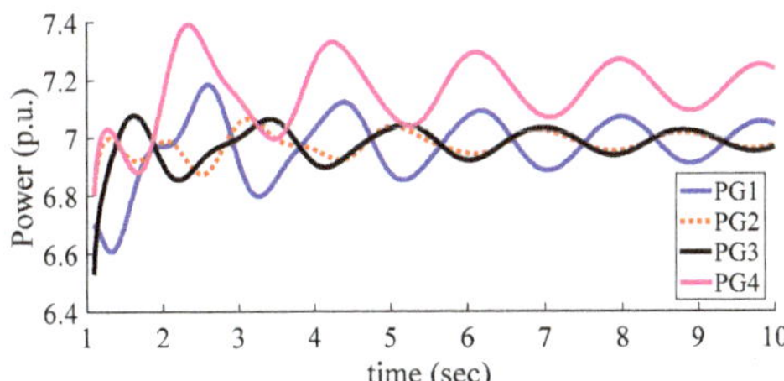

Figure 8. Power output of all generators related to Test Case 2.

Table 3. Comparison of Extracted Modes based on Active Power of Different Generators in Test Case 2; $\xi(-)$, $f_d(Hz)$.

G1		G2		G3		G4	
ξ	f_d	ξ	f_d	ξ	f_d	ξ	f_d
–1	0	–1	0	–1	0	–1	0
1	0	1	0	1	0	1	0
0.037	0.49	0.034	0.49	0.033	0.49	0.043	0.49
0.085	0.97	0.081	0.95	0.088	0.94	0.083	0.98

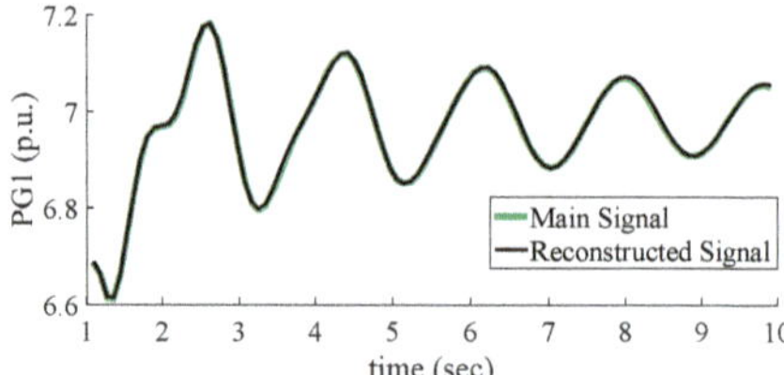

Figure 9. Comparison of main and reconstructed signals of PG1 in Test Case 2.

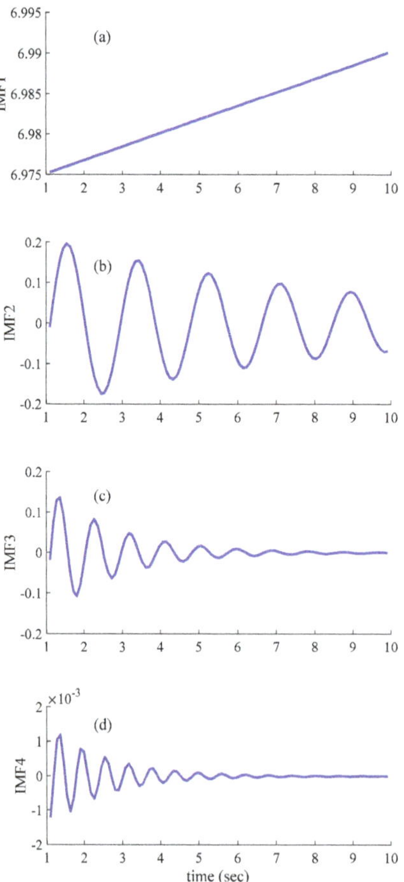

Figure 10. Extracted IMFs of PG1 in Test Case 2. (**a**–**d**) IMF1 to IMF4, respectively.

Finally, based on energy and phase angle of IMFs, contribution factor of each generator output power in each of the modes can be determined, as demonstrated in Table 4.

Table 4. Node contribution factor (NCF) matrix in Case 2.

Signal\Mode	Mode 1	Mode 2	Mode 3	Mod 4
PG1	4827.053	0.855	0.123	0.000
PG2	4825.077	0.120	0.059	0.000
PG3	4825.284	0.219	0.032	0.000
PG4	5090.697	1.031	0.000	0.072

The matrix in Table 4 implies that the most associated generation in oscillatory modes 1, 2, and 4 is generator G4, while mode 3 is mainly affected by generator G1.

3.3. Test Case 3: Interconnected Practical Power System

In this section, the proposed method of Sections 2.3–2.5 has been used to analyze measured signals in a practical transmission power system. The output powers of the five generators resampled at 10 Hz are used to demonstrate performance of the proposed method. All presented results in this section are generated by the interconnected power system WAMS-based oscillation detection application utilizing our proposed method of Sections 2.3–2.5. A general schematic of this application is demonstrated in Figure 11. The signals related to five generators (here they are named G1 to G5) associated with the area that are affected by the fault are depicted in Figure 11. As shown in Figure 11, the information of oscillatory modes are provided to the system operator in a format of a table. Columns 1 to 4 in the table represent real and imaginary part of each oscillator mode, and their damping ratio and frequency, respectively. Moreover, the fifth column demonstrates stability status of each mode. For example, as shown in Figure 11, the first mode is an unstable mode shown with red color. The next columns show type of modes based on their frequency, i.e., local or inter-area. According to Figure 11, mode 2 is an inter-area mode with frequency of 0.4 Hz, while Modes 3 and 4 are local modes with frequency of 0.94 and 4.4 Hz due to oscillation of Generator G1 output power.

Figure 11. A schematic of the power system oscillation detection application.

As an example, the comparison of the original and reconstructed generation G5 signal based on the Prony-based method is shown in Figure 12, which demonstrates an acceptable matching between two signals. The accuracy of the modes and consequently IMFs and NCFs can be concluded from this matching. Having IMFs associated with all five generation signals, a NCF matrix with five rows and four columns is calculated by our framework (see Section 2.4), as shown in Table 5. This table shows that, the most associated signal to modes 1, 2, and 4 is G5 and mode 3 is mainly affected by G1. Thus, for controlling and stabilizing the system against the unstable mode (mode 1), also increasing the damping of the poorly damped mode 4, G5 should be controlled. Also, observe that generation G3 is the most suitable actuator for controlling mode 3. As seen from Table 5, the energy of the mode 2 is less than mode 1, while the energy of modes 3 and 4 relative to mode 2 is insignificant. Note that as the energy of the other modes is negligible, our oscillation detection framework discards them from the representation of their parameters as well as IMFs and NCFs. For better representing the contribution of each generator in each mode, the values of the NCF matrix are normalized and plotted in Figure 13. Other important plots in the power system oscillation detection application are related to IMFs of the measured signals, which are not shown here for brevity. The plot is similar to those shown in Figures 6 and 10 in Test Case 1 and 2, respectively. These provide an increase intuitiveness in the proposed oscillation detection framework, beside NCF plots. It was observed in IMF plots that the main or dominant mode (mode 1) has a growing IMF as it is unstable, while the other modes (mode 2 to 5) have exponential sinusoidal IMFs which are damped after 40, 20, and 7 s, respectively.

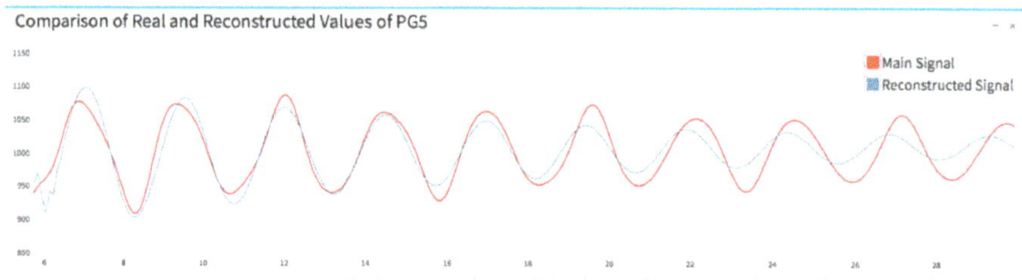

Figure 12. Comparison of real and reconstructed values of PG5 in Test Case 3.

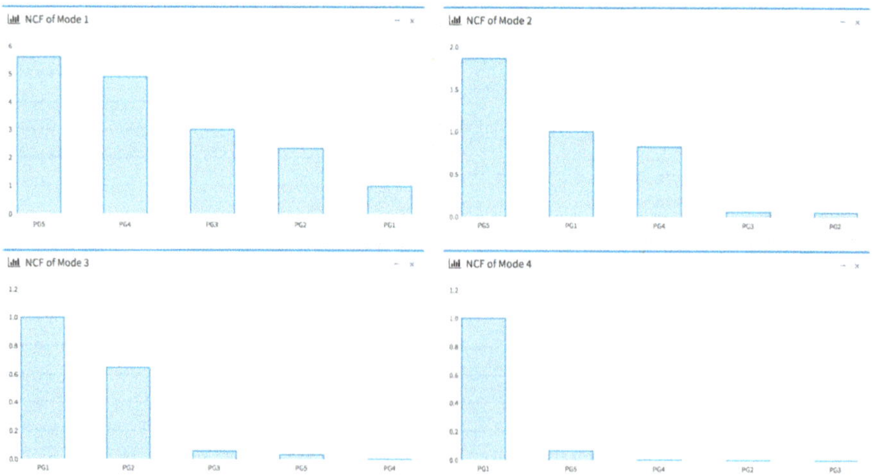

Figure 13. Contribution factors of the generators G1 to G5 of the studied third system in modes 1–4.

Table 5. NCF Matrix in Test Case 3.

Signal\Mode	Mode 1	Mode 2	Mode 3	Mode 4
PG1	0.437×10^8	1.349×10^5	4.662×10^4	2.28×10^3
PG2	1.023×10^8	0.138×10^5	507	399
PG3	1.314×10^8	0.163×10^5	140	497
PG4	2.140×10^8	1.764×10^5	136	2.85×10^3
PG5	2.448×10^8	4.631×10^5	1.380×10^4	3.91×10^3

3.4. Comparison with Related Methods

In this section, we show the superiority of the proposed method for oscillation analysis. Due to space limitations, the proposed augmented Prony-based oscillation analysis is compared with the well-known and adopted techniques such as HHT and conventional Prony methods. It is clear from Table 6 that the proposed method is superior to others in terms of indices such as IMF, NCF, computational complexity, EE, Gibbs, and mode-mixing. For instance, the proposed method can eliminate the worse effects of EE, Gibbs and mode-mixing where other methods cannot. Considering the computation burden as an index, the proposed method is superior to HHT in which the complexity order of the proposed augmented method and HHT are O (1) and O (n × m), respectively. Where n is the number of the desired modes and m is the number of iterations in EMD process.

Table 6. Comparison of the Proposed and Other Related Works.

Method\Index	IMF	NCF	Computational. Complexity	EE	Gibbs	Mode-mixing
HHT	√	√	O (nm)	×	×	×
Conventional Prony	×	×	-	-	-	-
Proposed method	√	√	O (1)	√	√	√

In order to further verify the proposed method, the results of the Test case 2 is compared with the results of PSAT software which is a MATLAB-based software, which can be used for small-signal stability analysis. Note that the analysis of PSAT is based on the detailed modeling of the generators, automatic voltage regulators (AVRs), power system stabilizers, PSSs, etc., whereas the proposed method is based on the synchrophasor measurements and the modeling of power system components is not needed. As seen from Table 7, the damping ratio and frequency of the dominant modes (Eig #21, Eig #22, Eig #13, and Eig #14) of the studied power system obtained by the detailed modeling in PSAT are comparable with the results of proposed method presented in Table 3. Both results, demonstrates that the dominant mode is an unstable mode. One of the capabilities of the PSAT toolbox is that it can determine the most-associated states related to each mode. The PSAT determines this using participation factors (PP) which be calculated using the elements of the state matrix A (if we consider the state equations as $\dot{x} = Ax + Bu$). Alternatively, in this paper we determine the most associated state variable corresponding to each oscillatory mode using the NCF index without the need to detail modeling of the system. As can be seen from Table 7, based on the PSAT results, the most associated generation in oscillatory modes 1, 2 is generator G4 which complies with the results obtained based on the NCF index in the proposed method.

Table 7. The Results of Test Case 2, Obtained by Detailed Modeling of Power System in PSAT Software.

Eigevalue	Most Associated States	Real Part	Imag. Part	Damping Ratio	freq
Eig #1	e2q_Syn_2	−36.498	0.000	1.000	0.000
Eig #2	e2q_Syn_4	−36.423	0.000	1.000	0.000
Eig #3	e2q_Syn_3	−34.055	0.000	1.000	0.000
Eig #4	e2q_Syn_1	−32.733	0.000	1.000	0.000
Eig #5	e2d_Syn_1	−30.466	0.000	1.000	0.000
Eig #6	e2d_Syn_2	−30.257	0.000	1.000	0.000
Eig #7	e2d_Syn_4	−23.408	0.000	1.000	0.000
Eig #8	e2d_Syn_3	−21.699	0.000	1.000	0.000
Eig #9	delta_Syn_1, omega_Syn_1	−0.562	6.798	0.082	1.082
Eig #10	delta_Syn_1, omega_Syn_1	−0.562	−6.798	0.082	1.082
Eig #11	omega_Syn_2, delta_Syn_2	−0.561	6.610	0.085	1.052
Eig #12	omega_Syn_2, delta_Syn_2	−0.561	−6.610	0.085	1.052
Eig #13	delta_Syn_4, omega_Syn_4	−0.126	3.457	0.036	0.550
Eig #14	delta_Syn_4, omega_Syn_4	−0.126	−3.457	0.036	0.550
Eig #15	e1d_Syn_1	−5.835	0.000	1.000	0.000
Eig #16	e1d_Syn_2	−5.787	0.000	1.000	0.000
Eig #17	e1d_Syn_4	−4.388	0.000	1.000	0.000
Eig #18	e1d_Syn_3	−3.495	0.000	1.000	0.000
Eig #19	e1q_Syn_1	−0.171	0.000	1.000	0.000
Eig #20	e1q_Syn_2	−0.179	0.000	1.000	0.000
Eig #21	omega_Syn_4	−0.069	0.000	1.000	0.000
Eig #22	delta_Syn_4	0.010	0.000	−1.000	0.000
Eig #23	delta_Syn_4	0.000	0.000	NaN	0.000
Eig#24	omega_Syn_4	0.000	0.000	NaN	0.000

4. Conclusions

In this paper, an online WAMS-based power system oscillation analysis framework was proposed. Features of the Prony method, such as simplicity in the implementation, self-verification, and robustness against the noise to estimate the modal information of a measured signal (i.e., its damping and frequency), nominated it for practical applications. The Prony method is then augmented here to be able to calculate IMFs and thus NCFs. IMFs provide an intuitive representation of the system modes and are mainly calculated using the iterative HHT-based methods. Our proposed framework calculates IMFs using an explicit mathematical formulation based on the augmented Prony method directly from PMU measurements. Therefore, it is simpler, faster and more accurate compared to those methodologies. The proposed Prony-based NCF index of this paper allowsthe identification of the most associated state variable to each mode accurately.

The simulation results on three test cases revealed that the proposed framework is able to determine IMFs and NCFs more accurately.

Author Contributions: All authors have contributed to the work. Conceptualization, Data curation, and Methodology: M.K.A., M.K., R.T.; Writing—review & editing: M.K.A., M.K., H.H.A., and P.S.

Funding: This research received no external funding.

Conflicts of Interest: The authors declare no conflict of interest.

References

1. Kundur, P. *Power System Stability and Control*; McGraw-Hill: New York, NY, USA, 1994; Volume 7.
2. Trudnowski, D.J.; Pierre, J.W.; Zhou, N.; Hauer, J.F.; Parashar, M. Performance of three mode-meter block-processing algorithms for automated dynamic stability assessment. *IEEE Trans. Power Syst.* **2008**, *23*, 680–690. [CrossRef]
3. Zhou, N.; Trudnowski, D.J.; Pierre, J.W.; Mittelstadt, W.A. Electromechanical mode online estimation using regularized robust RLS methods. *IEEE Trans. Power Syst.* **2008**, *23*, 1670–1680. [CrossRef]

4. Yazdanian, M.; Mehrizi-Sani, A.; Mojiri, M. Estimation of electromechanical oscillation parameters using an extended Kalman filter. *IEEE Trans. Power Syst.* **2015**, *30*, 2994–3002. [CrossRef]
5. Tripathy, P.; Srivastava, S.C.; Singh, S.N. A modified TLS-ESPRIT-based method for low-frequency mode identification in power systems utilizing synchrophasor measurements. *IEEE Trans. Power Syst.* **2011**, *26*, 719–727. [CrossRef]
6. Yang, D.C.; Rehtanz, C.; Li, Y. Analysis of low frequency oscillations using improved Hilbert-Huang transform. In Proceedings of the 2010 International Conference on Power System Technology (POWERCON), Hangzhou, China, 1–7 October 2010.
7. Barocio, E.; Pal, B.C.; Thornhill, N.F.; Messina, A.R. A dynamic mode decomposition framework for global power system oscillation analysis. *IEEE Trans. Power Syst.* **2015**, *30*, 2902–2912. [CrossRef]
8. Hu, X.; Peng, S.; Hwang, W.L. EMD revisited: A new understanding of the envelope and resolving the mode-mixing problem in AM-FM signals. *IEEE Trans. Signal Process.* **2012**, *60*, 1075–1086.
9. Yang, D.; Rehtanz, C.; Li, Y.; Tang, W. A novel method for analyzing dominant oscillation mode based on improved EMD and signal energy algorithm. *Sci. China Technol. Sci.* **2011**, *54*, 2493. [CrossRef]
10. Allen, A.J.; Singh, M.; Muljadi, E.; Santoso, S. Measurement-based investigation of inter-and intra-area effects of wind power plant integration. *Int. J. Elect. Power Energy Syst.* **2016**, *83*, 450–457. [CrossRef]
11. Turunen, J.; Thambirajah, J.; Larsson, M.; Pal, B.C.; Thornhill, N.F.; Haarla, L.C.; Hung, W.W.; Carter, A.M.; Rauhala, T. Comparison of three electromechanical oscillation damping estimation methods. *IEEE Trans. Power Syst.* **2011**, *26*, 2398–2407. [CrossRef]
12. Sarmadi, S.N.; Venkatasubramanian, V. Electromechanical mode estimation using recursive adaptive stochastic subspace identification. *IEEE Trans. Power Syst.* **2014**, *29*, 349–358. [CrossRef]
13. Khalid, H.M.; Peng, J.C.H. Tracking electromechanical oscillations: An enhanced maximum-likelihood based approach. *IEEE Trans. Power Syst.* **2016**, *31*, 1799–1808. [CrossRef]
14. Trudnowski, D.J.; Johnson, J.M.; Hauer, J.F. Making Prony analysis more accurate using multiple signals. *IEEE Trans. Power Syst.* **1999**, *14*, 226–231. [CrossRef]
15. Khazaei, J.; Fan, L.; Jiang, W.; Manjure, D. Distributed Prony analysis for real-world PMU data. *Elect. Power Syst. Res.* **2016**, *133*, 113–120. [CrossRef]
16. Zhao, S.; Loparo, K.A. Forward and Backward Extended Prony (FBEP) Method for Power System Small-Signal Stability Analysis. *IEEE Trans. Power Syst.* **2017**, *32*, 3618–3626. [CrossRef]
17. Ogrosky, H.R.; Stechmann, S.N.; Chen, N.; Majda, A.J. Singular Spectrum Analysis with Conditional Predictions for Real-Time State Estimation and Forecasting. *Geophys. Res. Lett.* **2019**, *46*, 1851–1860. [CrossRef]
18. Wu, Z.; Huang, N.E. Ensemble empirical mode decomposition: A noise-assisted data analysis method. *Adv. Adapt. Data Anal.* **2009**, *1*, 1–41. [CrossRef]

Article

Improving Microgrid Frequency Regulation Based on the Virtual Inertia Concept while Considering Communication System Delay

Gholam Ali Alizadeh [1], Tohid Rahimi [2,*], Mohsen Hasan Babayi Nozadian [2], Sanjeevikumar Padmanaban [3,*] and Zbigniew Leonowicz [4]

1 Department of Electrical and Computer Engineering, Faculty of Ghazi Tabatabai, Urmia Branch, Technical and Vocational University (TUV), Urmia 5716933959, Iran; Alizadeh_88@yahoo.com

2 Electrical and Computer Engineering Faculty, University of Tabriz, Tabriz 57734, Iran; m.hasanbabayi@tabrizu.ac.ir

3 Department of Energy Technology, Aalborg University, Esbjerg 6700, DK-9220 Aalborg, Denmark

4 Faculty of Electrical Engineering, Wroclaw University of Science and Technology, Wyb. Wyspianskiego 27, 50370 Wroclaw, Poland; zbigniew.leonowicz@pwr.edu.pl

* Correspondence: rahimitohid@tabrizu.ac.ir (T.R.); san@et.aau.dk (S.P.); Tel.: +98-9361996335 (T.R.)

Received: 6 April 2019; Accepted: 22 May 2019; Published: 26 May 2019

Abstract: Frequency stability is an important issue for the operation of islanded microgrids. Since the upstream grid does not support the islanded microgrids, the power control and frequency regulation encounter serious problems. By increasing the penetration of the renewable energy sources in microgrids, optimizing the parameters of the load frequency controller plays a great role in frequency stability, which is currently being investigated by researchers. The status of loads and generation sources are received by the control center of a microgrid via a communication system and the control center can regulate the output power of renewable energy sources and/or power storage devices. An inherent delay in the communication system or other parts like sensors sampling rates may lead microgrids to have unstable operation states. Reducing the delay in the communication system, as one of the main delay origins, can play an important role in improving fluctuation mitigation, which on the other hand increases the cost of communication system operation. In addition, application of ultra-capacitor banks, as a virtual inertial tool, can be considered as an effective solution to damp frequency oscillations. However, when the ultra-capacitor size is increased, the virtual inertia also increases, which in turn increases the costs. Therefore, it is essential to use a suitable optimization algorithm to determine the optimum parameters. In this paper, the communication system delay and ultra-capacitor size along with the parameters of the secondary controller are obtained by using a Non-dominated Sorting Genetic Algorithm II (NSGA-II) algorithm as well as by considering the costs. To cover frequency oscillations and the cost of microgrid operation, two fitness functions are defined. The frequency oscillations of the case study are investigated considering the stochastic behavior of the load and the output of the renewable energy sources.

Keywords: frequency stability; load frequency control; communication system delay; virtual inertia

---

1. Introduction

Microgrids are small power networks that are composed of loads, distributed generations, energy storage resources, telecommunication facilities and other infrastructure [1,2]. These networks are protected from damages resulting from the faults of the upstream network. The stochastic variations of generation units and load profiles [3,4] are common issues in these networks. The availability of different forms of energy storage in microgrids as well as their contribution to electricity markets at different time periods justify microgrids more economically. These networks are suitable for critical applications such as hospitals and military centers, which require high levels of reliability. Natural disasters and system faults events can rarely affect the performance of these systems [5].

With the increasing penetration of renewable energy technologies in power grids, these grids experience new challenges. In a microgrid, power sources are connected to the network by power electronic converters. These interfaces have very fast transient responses, which reduces the overall inertia of the system. Consequently, low-inertia microgrids are less stable against any electrical disturbances. The imbalance between supply and load leads to frequency fluctuation. This problem, in turn, is aggravated by the presence of renewable energy sources. Renewable energy sources have unpredictable and fluctuating generation modes [6–9]. Fortunately, the energy storage units are capable of functioning as a power source or as load, and therefore can help attenuate the frequency fluctuations by absorbing or supplying power [10–14].

For the stable operation of microgrids, balancing between demand and supply in real time is an essential step, so that the frequency can be constant in its acceptable range. This task can be fulfilled using the load-frequency control mechanism [15]. The microgrid central controller divides the demand power between the power supply sources, thus maintaining the frequency of the system in the acceptable range [16]. Kalantar and Mousavi have studied the dynamical behavior of a hybrid system, which consisted of a wind turbine, microturbine, solar array, and battery according to the genetic algorithm and fuzzy controller [17]. Gao et al. have introduced active topology in its experimental and laboratory sense to control another hybrid system, including a battery and fuel cell [18]. Nayyeri Pour et al. have implemented the PI frequency control method in a local hybrid network composed of photovoltaic, a wind turbine, fuel cell, and super-capacitor. In addition, one PI controller is considered for each source, which determines the reference power of each unit for frequency stabilization by receiving frequency deviation [19]. In reference [20], a fuzzy set and Harmonic Search Algorithm is proposed to adjust the PI controller coefficients in the microgrid in the presence of an electric drive. In reference [21], an optimized fuzzy control system is used to overcome the frequency fluctuations of the microgrid against load and generation uncertainties. One of the modern methods to smooth frequency oscillation and improve the stability of microgrids is virtual synchronous generators [22,23]. In this method, pseudo inertial forces are realized by controlling DGs inverters to emulate the behavior of synchronous generators virtually into microgrids [24]. A Battery/Ultra capacitor hybrid energy storage system is introduced in reference [25] to cover power management and virtual synchronous generator emulation goals. In reference [26], an adaptive control method based on a virtual inertia system with MPC has been presented for frequency oscillation damping by emulating virtual inertia into the microgrid. The virtual inertia control is molded by a first-order transfer function in reference [26]. In reference [27], ultra-capacitor systems are controlled in a way that increases the virtual inertia of the studied microgrid. However, it is important to optimize the ultra-capacitor bank size to avoid extra costs.

On the other hand, with the development of computing and telecommunication systems, communication media are being used to exchange power between the central processing unit and the loads. However, with the increasing penetration of telecommunication networks in the control tasks, the delay between secondary controller and power resources cannot be ignored. Researchers have investigated the stability of the frequency control system in the presence of a time delay in several systems [27–29]. With decreasing communication delay, the cost of the communication infrastructure increases, however the frequency stability remains in a suitable state.

As discussed above, improving the inertia of microgrid and reducing the time delay of the communication system are effective methods to overcome frequency oscillations. However, this approach increases the operation cost of a microgrid. It is essential to select optimum values of time delay and Ultra capacitor size to have acceptable frequency oscillation of a microgrid and operation costs. To sum up, the challenges that discussed in this paper are given as follows:

- To reduce the installation costs, the size of the installed equipment should be minimal, so that the system will not face high costs during its long operation time.

- To enhance the frequency behavior of the microgrid, communicational infrastructure with minimum delays is required to update information quickly.

With considering the challenges that discussed above, this paper aim is determining optimal virtual inertia, comunication delay time and frequency control parameters simultaneously that are not investigated in privious papers. The proposed strategy guarantees frequency stability and decreases the cost of the microgrid's infrastructure. A Minimum ultra-capacitor size and communication delay value are selected in the proposed approach to decrease the cost of the microgrid's infrastructure. Also, the stable and unstable regions for different time delay and virtual inertia values are derived.

Multi-objective optimization methods are used to determine the parameters used in the load-frequency strategies to optimize the several objectives discussed previously [30]. The NSGA-II algorithm is a popular and justified optimizing algorithm [31,32]. Therefore, The NSGA-II algorithm is used in this paper to address the ultra-capacitor size as virtual inertia and the delay of the telecommunications system.

2. Case Study

Here, a microgrid system, which is isolated from the main grid, is considered for this case study. The islanded system consists of power generation, storage units, and telecommunication configurations. Diesel Generator Unit, Fuel Cell Unit, Solar Unit, and Wind Unit are responsible for power generation. The electrolyze unit is also use to absorb extra power. Ultra-capacitors are additionally used to generate and/or absorb power quickly in order to increase the virtual inertia of the network.

The central controller needs frequency information to appropriately order the microgrid power generation and absorption units. As discussed in reference [33], frequency and power information are to be sent or received with a specific time delay by a telecommunications platform, which is very effective for the stability of the microgrid. The communication delay with other delay sources are defined as a measurement delay in reference [34] and applied for primary and secondary controllers. The microgrid mentioned before is shown in Figure 1. Δf is the frequency deviation and input signal of the controller, and the changes in the power generation and storage units as the output of this system. M represents the total inertia of a microgrid. When a difference between the generation and consumption occurs because of the load along with power generation increasing/decreasing, the frequency deviation of the microgrid depends on the inertia value and the load demand. It is assumed that the solar unit has variable and uncontrollable power. The central controller block sets a PI controller output for each power unit. The central controller unit block receives the network frequency information with a delay. The power and frequency information are passed through with a telecommunication system delay. The Diesel Generator Unit, Fuel Cell Unit, and electrolyze unit are equipped with primary and secondary frequency controllers. The Ultra-Capacitor unit participates in primary frequency control using a frequency signal. The parameters of this microgrid and power units are given in Table 1.

Table 1. Microgrid Constant Parameters.

Parameter	Value	Parameter	Value
T_{uc}	0.2 (s)	T_{FC}	4 (s)
H	5 (s)	T_{AC}	0.2 (s)
D	0.012 (pu/Hz)	T_{dt}	20 (s)
T_{dg}	2 (s)	f_{base}	50 Hz

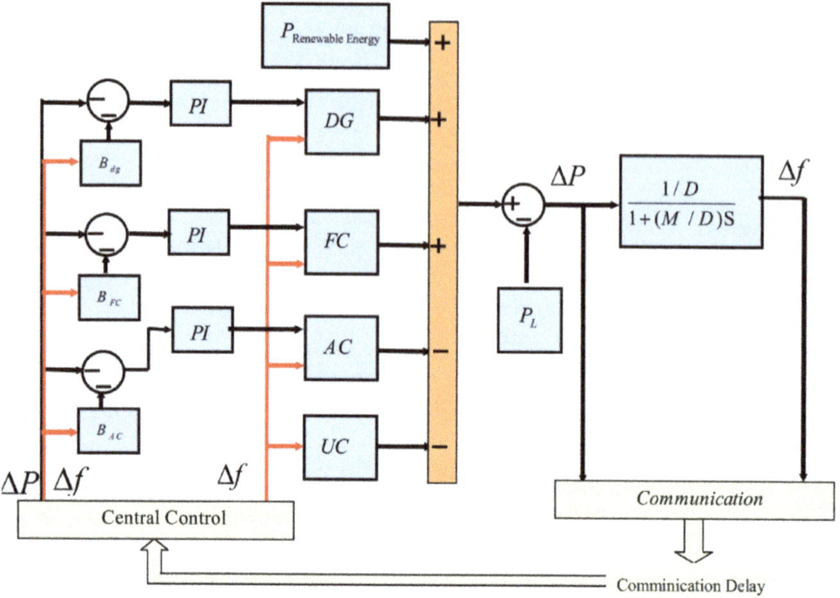

Figure 1. Illustration of the Control Units of the Studied Microgrid.

2.1. Modeling the Power Source of Diesel Generator

The diesel generators are used to compensate the power required by the microgrid by using a governor and a speed drop system. The governor controls the fuel input of the fuel engine by adjusting the fuel throttle mechanism. The gasoline or gas engine acts as a turbine and moves a synchronous generator. The diesel generator governor can be modeled by a first-order conversion function that is described in Equation (1).

$$G_{dgs}(s) = \frac{1}{1 + sT_{dgs}} \tag{1}$$

Similarly, the turbine of this generator is also modeled by the first order conversion function.

$$G_{dt}(s) = \frac{1}{1 + sT_{dt}} \tag{2}$$

Due to the series performance of the mechanical and electrical unit of the diesel generator, the general conversion function is as follows.

$$G_{dgt}(s) = \frac{1}{1 + sT_{dgs}} \times \frac{1}{1 + sT_{dt}} \tag{3}$$

The control commend to the diesel generator cannot be transferred to its output immediately because of the inherent delay of mechanical systems and generation rate constraint (GRCs). Therefore, these limitations are also applied to the Diesel Dynamic Model. The complete dynamic model of the diesel generator is shown in Figure 2.

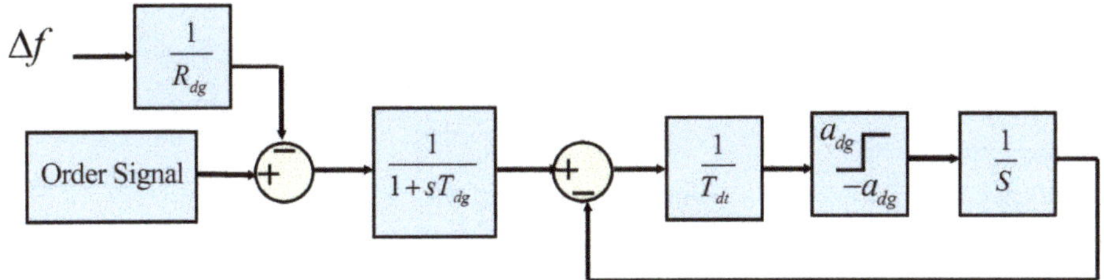

Figure 2. Dynamic Model of the Diesel Generator [33].

2.2. Fuel Cell

The fuel cell is one of the static power generation equipment that converts the chemical energy stored in hydrogen into DC electrical energy. When the generated power of the renewable energy source is less than the power required for electric loads or the consumption of loads at their peak, the fuel cell begins to generate power by injecting hydrogen stored in its hydrogen tank. Since the ultimate performance of this system is similar to the voltage source inverter, the transfer function of this power source will be in the form of the first-order model as follows:

$$G_{fc}(s) = \frac{1}{1 + sT_{fc}} \tag{4}$$

The generated power and its increase rate limit are also applied for the fuel cell. The dynamic model of a fuel cell is illustrated in Figure 3.

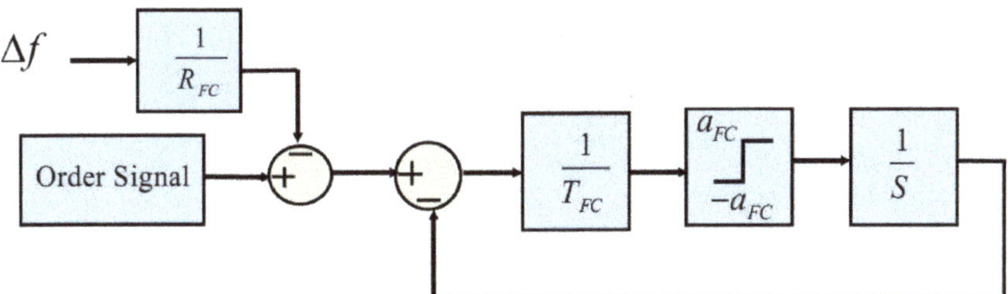

Figure 3. Dynamic Model of the Fuel Cell [33].

2.3. Electrolyze Unit

An electrolyze unit is a form of equipment that stores the extra energy of renewable energy sources. Water is decomposed into hydrogen and oxygen by passing the current between two separated electrodes. The hydrogen produced in the chemical reaction is stored in a tank and used as fuel for a fuel cell to meet load demand.

The water electrolyzes unit acts as a DC voltage source and requires a VSC inverter to be connected to a microgrid. Consequently, the transfer function will be as follows:

$$G_{AE}(s) = \frac{K_{AE}}{1 + sT_{AE}} \tag{5}$$

The power generation rate limits are also considered in the water electrolyze model. Figure 4 shows the dynamic model of the electrolyze unit.

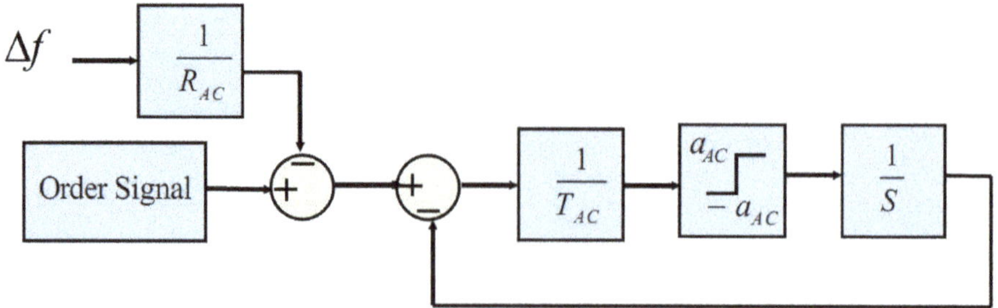

Figure 4. Dynamic Model of the Electrolyzer [33].

2.4. Model of Renewable Energy Sources

In case of renewable energies, the MPPT strategy is implemented, so controllability will not be possible for wind/PV power generation [6,7]. These systems are not participating in frequency control.

2.5. Ultra-Capacitor

The Ultra-capacitors plays an important role in increasing the virtual inertia of the network, either by power injection or by absorption. The network exchange power of the Ultra-capacitor is directly related to the derivation of frequency fluctuations. These units have very little delay in response to frequency oscillations. Thus, the dynamic model of ultra-capacitors can be illustrated in Figure 5 [27].

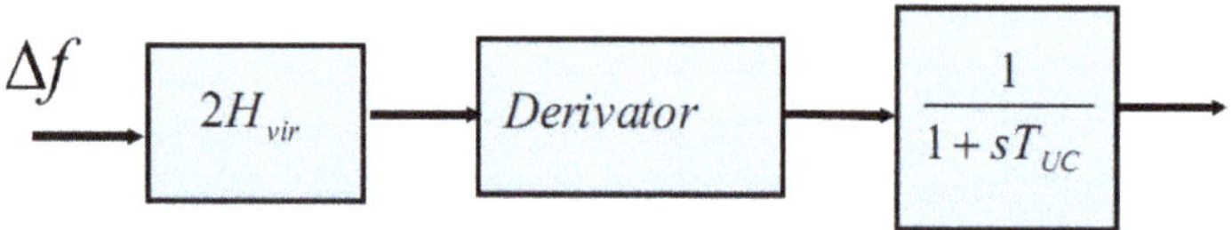

Figure 5. Dynamic Model of Ultra-capacitor.

2.6. Telecommunication System Delay

The telecommunication system is used to transfer microgrid information to the central controller. In reference [34], a first order transfer function is considered for measurement of a delay that covers the communication delay. Therefore, for the simplification of analyses, all delays in communication units modeled as the first-order function in this paper, which is expressed in Equation (6). It should be noted that any other time delay model can be used and the optimizing algorithm can determine the optimum parameters of the selected model.

$$G_{Commu}(s) = \frac{1}{1 + sT_{commu}} \tag{6}$$

2.7. Optimization Algorithm

The NSGA-II algorithm is used in this paper. The first goal is reducing the frequency fluctuations of the microgrid. Equation (7) is considered for this aim. The algorithm attempts to minimize (7) so the frequency oscillation will be in a minimum state. In addition, the second goal is determining the minimum ultra-capacitor size and the delay value of telecommunication systems to reduce the microgrid infrastructure cost. For a low ultra-capacitor size and the high delay value of telecommunication systems, the operation cost will be low. These two goals are expressed in two separate objective functions as below:

$$IAE = \sum_{i=1}^{N} |\Delta f|_i + |f|_i. \tag{7}$$

$$CostF = \left(H_{Vir\_average} - H\right) + \frac{T_{d\_avarge}}{T_d} = H_{Vir} + \frac{T_{d\_avarge}}{T_d} \tag{8}$$

In the above Equations, i.e., (7) and (8), Δf stands for frequency fluctuations, H_{Vir} is the virtual inertia, T_d is the delay of telecommunication systems, T_{d_avarge} is an average delay of telecommunication systems, and N is the number of samples in the simulation process.

The accuracy of resources and microgrid frequency response models has been investigated in the previous literature. These models have been used for the optimization process. The upper and lower limits of the decision variables of the optimization problem are listed in Table 2.

Table 2. The parameters limitation ranges.

Parameter	Limitation	Parameter	Limitation
$H_{Vir\_average}$	5–8	B_{fc}	0.05–0.3
kp_{dg}	0.1–1	R_{fc}	10–20
ki_{dg}	0.001–0.005	kp_{ac}	60–80
B_{dg}	0.07–0.2	ki_{ac}	0.01–0.5
R_{dg}	10–20	B_{ac}	0.1–3
k_{pfc}	0.01–0.3	T_{ac}	10–20
ki_{fc}	0.001–0.005	T_d	1–10

The NSGA-II algorithm, like the conventional genetic algorithm, uses mutation and crossover operators at each stage. The algorithm uses repetitive loops to obtain the optimum population. The set of solutions for the right choice must first be ranked and each solution with fewer ratings is a better solution. This algorithm also begins its operation with the initial population, based on the values of the fitness functions. The solutions are ranked with the use of the mechanism of crowding distance and dominance, and based on this, a new population is generated with crossover and mutation functions [35]. As long as the two objective functions reach convergence, the algorithm should be continued. Figure 6 shows the convergence of these two conflicting functions in the the optimization algorithm steps.The multi-objective optimization is implemented in MATLAB/SIMULINK software

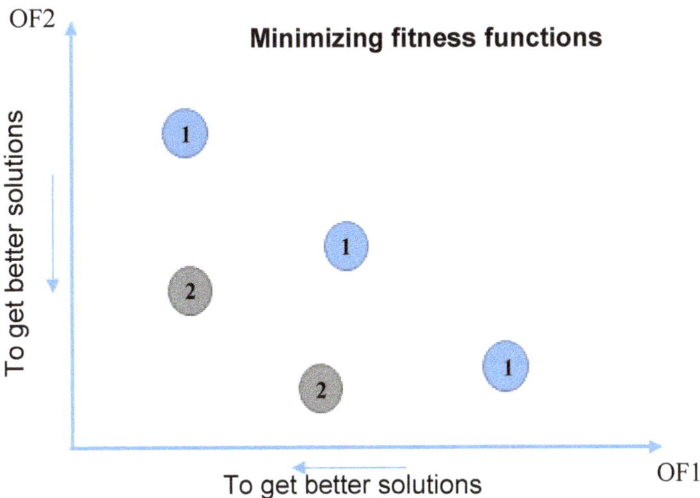

Figure 6. Pareto-based optimum solutions selection in the multi-objective optimization process.

3. Simulation Results

The frequency control system implemented in the MATLAB / SIMULINK software. The system parameters are obtained using the NSGA-II algorithm. The NSGA-II algorithm is used to optimize two conflict objectives, i.e. operation cost reduction and frequency oscillation reduction. The optimizing processes are showed as Pareto curves. The final Pareto curve is shown in Figure 7. The selected solution presented in the Pareto curve is considered an equally good solution. Low frequency oscillation with low operation costs can be derived in the selected point. The parameters values for the selected solution listed in Table 3. The optimum and non-optimum values are listed in Table 3. The non-optimum parameters were obtained by the trial and error method, reducing the infrastructure costs. The system was simulated in three scenarios:

- ✓ Scenario I: In the first scenario, the performance of the microgrid was simulated through the application of a telecommunication networks delay with non-optimum parameter values.
- ✓ Scenario II: In the second scenario, the performance of the microgrid with the optimum control parameters, virtual inertia and telecommunication systems delay values is verified through simulation setups.
- ✓ Scenario III: In the third scenario, the delay of the telecommunication system is not considered while the optimum parameters are applied in the simulated system.

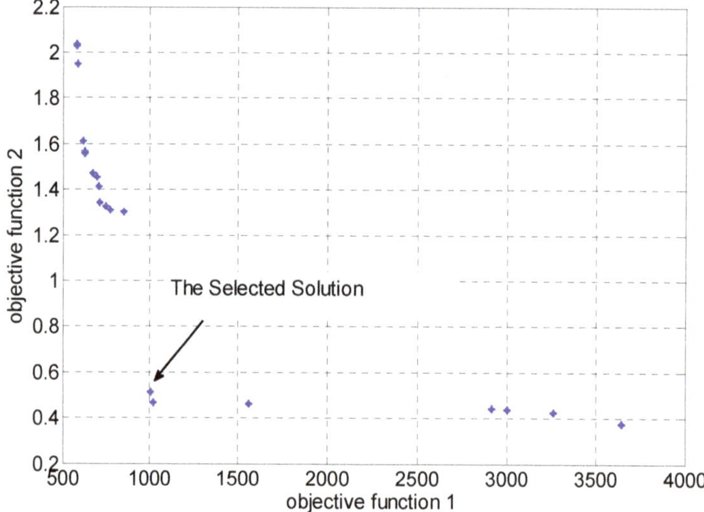

Figure 7. Frontiers of Pareto of the two objective functions drawn from NSGA-II optimization.

Table 3. Optimum and non-optimum parameters of the control system, ultra-capacitor size, and telecommunications system delay values.

Non-Optimum Parameters		Optimum Parameters	
Parameter	Value	Parameter	Value
$Hvir$	1	$Hvir$	0.2005
kp_{dg}	0.244	kp_{dg}	0.7167
ki_{dg}	0.0032	ki_{dg}	0.0024
B_{dg}	0.1386	B_{dg}	0.1687
R_{dg}	19.93	R_{dg}	17.084
kp_{fc}	0.0298	kp_{fc}	0.0764
ki_{fc}	0.0043	ki_{fc}	0.0025
B_{fc}	0.1878	B_{fc}	0.2338
R_{fc}	17.09	R_{fc}	12.7212
kp_{ac}	57.7635	kp_{ac}	75.3
ki_{ac}	7.08	ki_{ac}	0.4
B_{ac}	0.5923	B_{ac}	0.1
R_{ac}	28.1	R_{ac}	16.1
T_d	3	T_d	4.8
$CostF$	1.2	$CostF$	0.513

These scenarios show the effectiveness of the selected solution in overcoming frequency oscillation. In all the scenarios, the load and renewable energy source step changes are considered in the same profile. The magnitude of the load demand is equal to 1/1 per unit. When the simulation time reaches 1500 (s), it will reach 0.9 per unit. In contrast, the generation of renewable energy units is equal to one per unit. The frequency behavior of the microgrid in this section is shown in Figure 8. The microgrid frequency fluctuation has decreased with the application of optimum parameters and can be clearly seen in Figure 8. Meanwhile, the telecommunication system delay increases the undershoot and settling time of the frequency fluctuation. Up to t = 1500 s, diesel generator and fuel cell are expected to generate power, and after this instance the electrolysis unit is expected to absorb unbalanced power. The dynamic responses of the power sources and the electrolyze unit, while applying the optimum parameters, can be investigated using simulation results. Output powers of the source for the diesel generator, the fuel cell, and the electrolyze unit are shown in Figures 9–11, respectively. As can be seen from these figures, the electrolysis unit and diesel generator have the fastest and the slowest dynamic behavior responses, respectively. In the initial simulation time, the fuel cell injects power to grid with a high rate to compensate frequency drooping. Up to t = 1500 s, in steady state condition, the diesel generator and fuel cell contribute in power balancing and frequency stability.

Due to the high penetration of renewable energy sources, the actual inertia of the microgrid systems is small. This phenomenon shows its effect on steady frequency oscillation and high overshoot and undershoots in microgrid frequency responses. By increasing the communication delay, the stability margin of the microgrid can be improved. By increasing the microgrid inertia by increasing the virtual inertia value, the microgrid frequency is made less sensitive to the increasing commendation time delay. The studied microgrid behavior, for a different time delay value, is stable under the special condition that is shown in Figure 12. Two areas are distinguished in this figure. For a given H_{vir} value, if the time delay value is in the blue area, the system will be stable. The maximum virtual inertia margin that guarantees the stability is shown in Figure 13. To derive Figures 10 and 11, the conventional repetition process is used. The virtual inertia and time delay step changes are considered as being 0.5 Hz and 0.025 s, respectively.

With respect to general load and power generation profile, shown in Figure 14, the frequency behaviour of the grid while applying the variable profiles of the load power and power generated by renewable energy sources is shown in Figure 14. The frequency behavior of the grid with optimum parameters (Figure 15) shows that the proposed strategy is more effective than the non-optimum based controller in damping frequency oscillation.

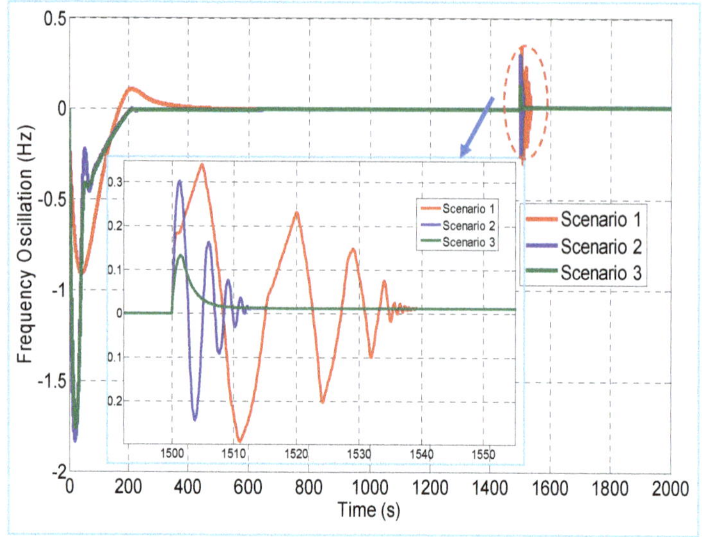

Figure 8. Comparison of the microgrid responses in terms of frequency in three scenarios.

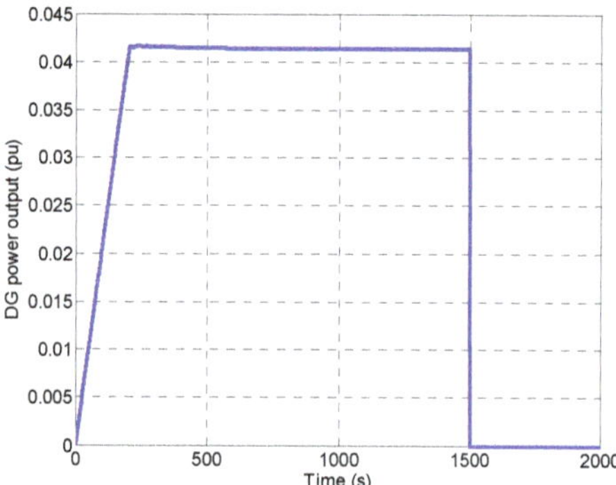

Figure 9. The diesel Generator output power.

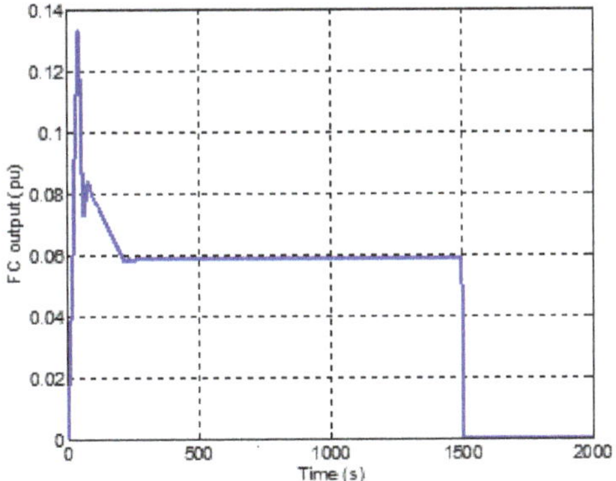

Figure 10. The fuel cell output power.

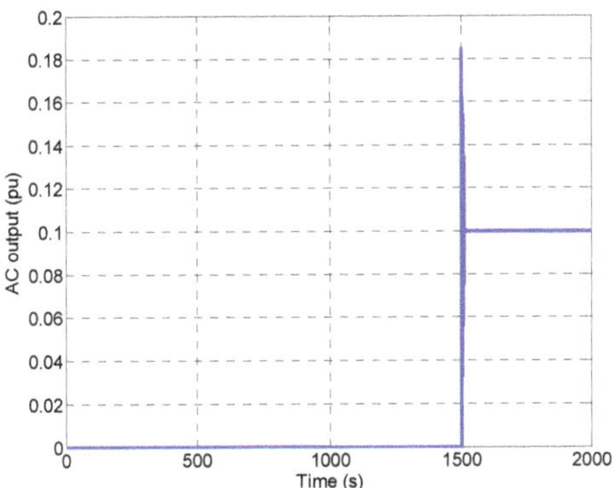

Figure 11. The electrolyzer output power.

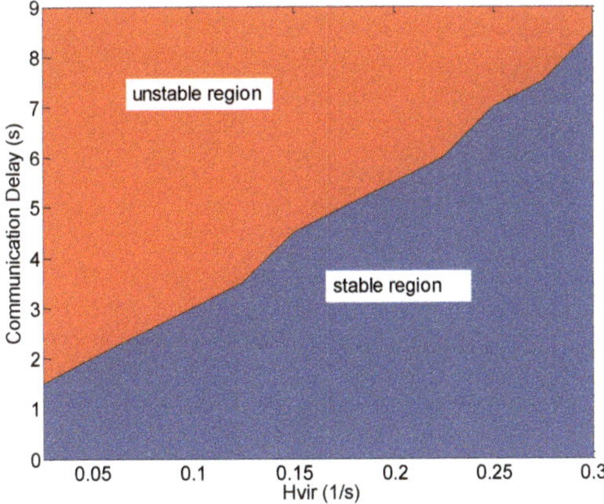

Figure 12. The stable and unstable regains for the studied microgrid.

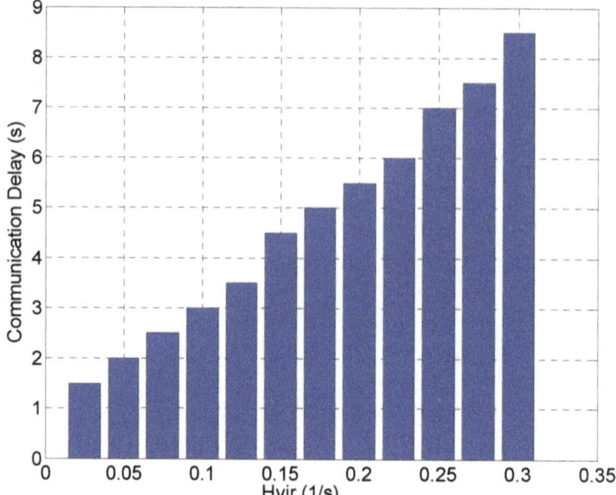

Figure 13. The maximum acceptable communication delay for given virtual values.

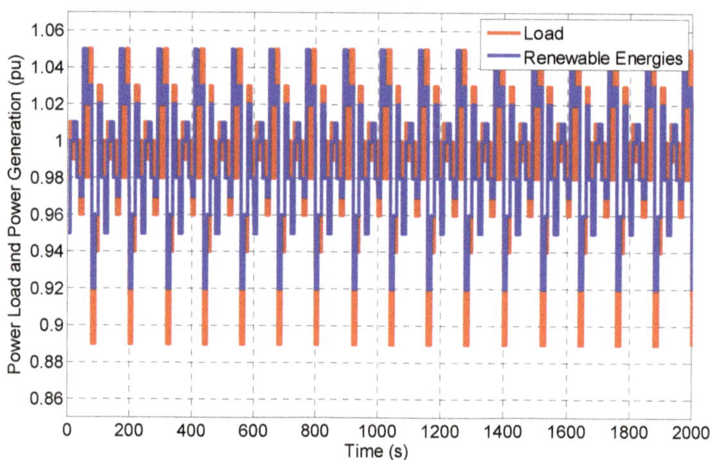

Figure 14. The renewable energy output and load demand powers.

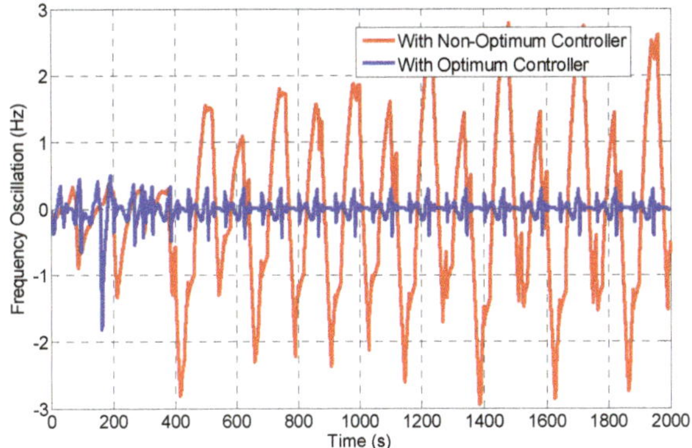

Figure 15. Frequency deviation with optimum and non-optimum controllers.

In real grids, the output power of the renewable energy sources and load power consumption are stochastic. In this paper, the Gaussian distributions are assumed for the random output of the renewable energy sources and load profile, which expressed mathematically in Equations (9) and (10), respectively.

$$g_1(x) = \frac{1}{\sqrt{0.0005}} \exp(\frac{-1}{2}\left(\frac{x-1}{\sqrt{0.0005}}\right)^2) \tag{9}$$

$$g_2(x) = \frac{1}{\sqrt{0.001}} \exp(\frac{-1}{2}\left(\frac{x-1}{\sqrt{0.001}}\right)^2) \tag{10}$$

The frequency behavior of a microgrid while applying the specific uncertainties, optimum and non-optimum parameters is simulated and the results of these simulations are shown in Figure 16. The optimum-based controller can overcome power-unbalancing events and stabilize the frequency.

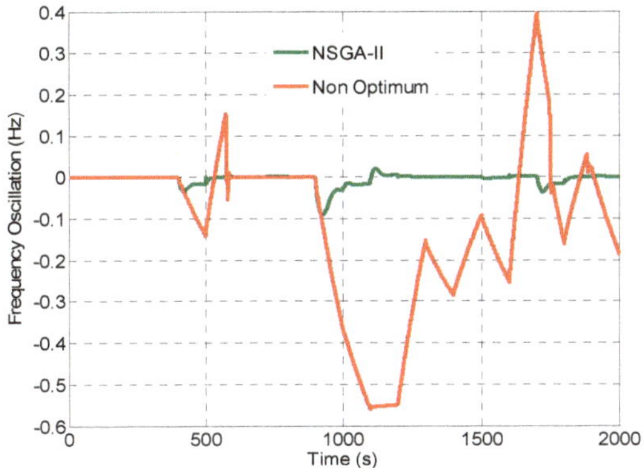

Figure 16. Frequency deviation under uncertainties.

4. Conclusions

A microgrid with controllable and uncontrollable sources, including solar, wind, a diesel generator, fuel cell, and an electrolyze unit, was investigated in this paper. The frequency stability of the grid was investigated while considering the ultra-capacitor unit and time delay of communication systems. The communication delay decreases the system function's ability to control the frequency. In this paper, tuning the virtual inertia constant and the time delay value in the communication together with the parameters of the load-frequency controllers of the power sources was considered to improve the frequency stability of low inertia microgrids with the presence of the time delay in the communication network. In fact, the best way to tune these unknown parameters has been determined as a multi-objective optimization problem (NSGA-II algorithm), which considers both frequency stability and economic concerns. Reducing delay in telecommunication systems and increasing virtual inertia values, while applying the ultra-capacitor, can improve the damping of frequency fluctuations; however, it also increases the cost of grid operation. With the tuning of the control system parameters, time delay value and the virtual inertia value, the aforementioned challenges are solved. In the studied microgrid, the simulation was performed in MATLAB/Simulink numerical software. Investigation results prove the effectiveness of this network strategy to overcome frequency fluctuations.

Author Contributions: All authors involved equally and contributed for the final.

Funding: This research received no external funding.

Conflicts of Interest: The authors declare no conflict of interest.

References

1. Hatziargyriou, N. (Ed.) *Microgrids: Architectures and Control*; John Wiley & Sons: Hoboken, NJ, USA, 2014.
2. Liu, W.; Gu, W.; Sheng, W.; Meng, X.; Wu, Z.; Chen, W. Decentralized multi-agent system-based cooperative frequency control for autonomous microgrids with communication constraints. *IEEE Trans. Sustain. Energy* **2014**, *5*, 446–456. [CrossRef]
3. Abdel-Hamed, A.M.; Ellissy, A.E.E.K.; El-Wakeel, A.S.; Abdelaziz, A.Y. Optimized control scheme for frequency/power regulation of microgrid for fault tolerant operation. *Electr. Power Compon. Syst.* **2016**, *44*, 1429–1440. [CrossRef]
4. Venkataramanan, G.; Marnay, C. A larger role for microgrids. *IEEE Power Energy Mag.* **2008**, *6*, 78–82. [CrossRef]
5. Chen, C.; Wang, J.; Qiu, F.; Zhao, D. Resilient distribution system by microgrids formation after natural disasters. *IEEE Trans. Smart Grid* **2016**, *7*, 958–966. [CrossRef]
6. Tani, A.; Camara, M.B.; Dakyo, B. Energy management in the decentralized generation systems based on renewable energy—Ultracapacitors and battery to compensate the wind/load power fluctuations. *IEEE Trans. Ind. Appl.* **2015**, *51*, 1817–1827. [CrossRef]
7. Ahmed, N.A.; Miyatake, M.; Al-Othman, A.K. Power fluctuations suppression of stand-alone hybrid generation combining solar photovoltaic/wind turbine and fuel cell systems. *Energy Convers. Manag.* **2008**, *49*, 2711–2719. [CrossRef]
8. Chen, Z.; Spooner, E. Grid power quality with variable speed wind turbines. *IEEE Trans. Energy Convers.* **2001**, *16*, 148–154. [CrossRef]
9. Liang, X. Emerging power quality challenges due to integration of renewable energy sources. *IEEE Trans. Ind. Appl.* **2017**, *53*, 855–866. [CrossRef]
10. Díaz-González, F.; Sumper, A.; Gomis-Bellmunt, O.; Villafáfila-Robles, R. A review of energy storage technologies for wind power applications. *Renew. Sustain. Energy Rev.* **2012**, *16*, 2154–2171. [CrossRef]
11. Shim, J.W.; Cho, Y.; Kim, S.J.; Min, S.W.; Hur, K. Synergistic control of SMES and battery energy storage for enabling dispatchability of renewable energy sources. *IEEE Trans. Appl. Supercond.* **2013**, *23*, 5701205. [CrossRef]
12. Onar, O.C.; Uzunoglu, M.; Alam, M.S. Dynamic modeling, design and simulation of a wind/fuel cell/ultra-capacitor-based hybrid power generation system. *J. Power Sources* **2006**, *161*, 707–722. [CrossRef]
13. Jing, W.; Lai, C.H.; Wong, W.S.; Wong, M.D. Dynamic power allocation of battery-supercapacitor hybrid energy storage for standalone PV microgrid applications. *Sustain. Energy Technol. Assess.* **2017**, *22*, 55–64. [CrossRef]
14. Chandrakala, V.; Sukumar, B.; Sankaranarayanan, K. Load frequency control of multi-source multi-area hydro thermal system using flexible alternating current transmission system devices. *Electr. Power Compon. Syst.* **2014**, *42*, 927–934. [CrossRef]

15. Arani, M.F.M.; Mohamed, Y.A.R.I. Cooperative control of wind power generator and electric vehicles for microgrid primary frequency regulation. *IEEE Trans. Smart Grid* **2018**, *9*, 5677–5686. [CrossRef]
16. Liu, Y.; You, S.; Tan, J.; Zhang, Y.; Liu, Y. Frequency Response Assessment and Enhancement of the US Power Grids toward Extra-High Photovoltaic Generation Penetrations—An Industry Perspective. *IEEE Trans. Power Syst.* **2018**, *33*, 3438–3449. [CrossRef]
17. Kalantar, M. Dynamic behavior of a stand-alone hybrid power generation system of wind turbine, microturbine, solar array and battery storage. *Appl. Energy* **2010**, *87*, 3051–3064. [CrossRef]
18. Gao, L.; Jiang, Z.; Dougal, R.A. An actively controlled fuel cell/battery hybrid to meet pulsed power demands. *J. Power Sources* **2004**, *130*, 202–207. [CrossRef]
19. Nayeripour, M.; Hoseintabar, M.; Niknam, T. Frequency deviation control by coordination control of FC and double-layer capacitor in an autonomous hybrid renewable energy power generation system. *Renew. Energy* **2011**, *36*, 1741–1746. [CrossRef]
20. Khooban, M.H.; Niknam, T.; Blaabjerg, F.; Dragičević, T. A new load frequency control strategy for micro-grids with considering electrical vehicles. *Electr. Power Syst. Res.* **2017**, *143*, 585–598. [CrossRef]
21. Sigrist, L.; Egido, I.; Miguélez, E.L.; Rouco, L. Sizing and controller setting of ultracapacitors for frequency stability enhancement of small isolated power systems. *IEEE Trans. Power Syst.* **2015**, *30*, 2130–2138. [CrossRef]
22. Yang, T.; Zhang, Y.; Wang, Z.; Pen, H. Secondary frequency stochastic optimal control in independent microgrids with virtual synchronous generator-controlled energy storage systems. *Energies* **2018**, *11*, 2388. [CrossRef]
23. Hammad, E.; Farraj, A.; Kundur, D. On Effective Virtual Inertia of Storage-Based Distributed Control for Transient Stability. *IEEE Trans. Smart Grid* **2019**. Accepted for publications. [CrossRef]
24. Alipoor, J.; Miura, Y.; Ise, T. Power system stabilization using virtual synchronous generator with alternating moment of inertia. *IEEE J. Emerg. Sel. Top. Power Electron.* **2015**, *3*, 451–458. [CrossRef]
25. Kerdphol, T.; Rahman, F.; Mitani, Y.; Hongesombut, K.; Küfeoğlu, S. Virtual inertia control-based model predictive control for microgrid frequency stabilization considering high renewable energy integration. *Sustainability* **2017**, *9*, 773. [CrossRef]
26. Fang, J.; Li, H.; Tang, Y.; Blaabjerg, F. Distributed power system virtual inertia implemented by grid-connected power converters. *IEEE Trans. Power Electron.* **2018**, *33*, 8488–8499. [CrossRef]
27. Jiang, L.; Yao, W.; Wu, Q.H.; Wen, J.Y.; Cheng, S.J. Delay-dependent stability for load frequency control with constant and time-varying delays. *IEEE Trans. Power Syst.* **2012**, *27*, 932–941. [CrossRef]
28. Liu, S.; Wang, X.; Liu, P.X. Impact of Communication Delays on Secondary Frequency Control in an Islanded Microgrid. *IEEE Trans. Ind. Electron.* **2015**, *62*, 2021–2031. [CrossRef]
29. Singh, V.P.; Kishor, N.; Samuel, P. Load frequency control with communication topology changes in smart grid. *IEEE Trans. Ind. Inform.* **2016**, *12*, 1943–1952. [CrossRef]
30. Wang, H.; Zeng, G.; Dai, Y.; Bi, D.; Sun, J.; Xie, X. Design of a Fractional Order Frequency PID Controller for an Islanded Microgrid: A Multi-Objective Extremal Optimization Method. *Energies* **2017**, *10*, 1502. [CrossRef]
31. Chaine, S.; Tripathy, M.; Satpathy, S. NSGA-II based optimal control scheme of wind thermal power system for improvement of frequency regulation characteristics. *Ain Shams Eng. J.* **2015**, *6*, 851–863. [CrossRef]
32. Kamjoo, A.; Maheri, A.; Dizqah, A.M.; Putrus, G.A. Multi-objective design under uncertainties of hybrid renewable energy system using NSGA-II and chance constrained programming. *Int. J. Electr. Power Energy Syst.* **2016**, *74*, 187–194. [CrossRef]
33. Mishra, S.; Mallesham, G.; Jha, A.N. Design of controller and communication for frequency regulation of a smart microgrid. *IET Renew. Power Gener.* **2012**, *6*, 248–258. [CrossRef]
34. Abazari, A.; Monsef, H.; Wu, B. Coordination strategies of distributed energy resources including FESS, DEG, FC and WTG in load frequency control (LFC) scheme of hybrid isolated micro-grid. *Int. J. Electr. Power Energy Syst.* **2019**, *1*, 535–547. [CrossRef]
35. Deb, K.; Pratap, A.; Agarwal, S.; Meyarivan, T.A.M.T. A fast and elitist multiobjective genetic algorithm: NSGA-II. *IEEE Trans. Evol. Comput.* **2002**, *6*, 182–197. [CrossRef]

Article

Black-Box Behavioral Modeling of Voltage and Frequency Response Characteristic for Islanded Microgrid

Yong Shi, Dong Xu *, Jianhui Su, Ning Liu, Hongru Yu and Huadian Xu

Anhui New Energy Utilization and Energy Saving Laboratory, School of Electrical Engineering and Automation, Hefei University of Technology, Hefei 230009, China; shiyong@hfut.edu.cn (Y.S.); su_chen@hfut.edu.cn (J.S.); Ning.Liu@unb.ca (N.L.); 2017010029@mail.hfut.edu.cn (H.Y.); xuhuadian@mail.hfut.edu.cn (H.X.)

* Correspondence: 2017110310@mail.hfut.edu.cn; Tel.: +86-0551-6290-4042

Received: 8 May 2019; Accepted: 27 May 2019; Published: 29 May 2019

Abstract: The voltage and frequency response model of microgrid is significant for its application in the design of secondary voltage frequency controller and system stability analysis. However, most models developed for this aspect are complex in structure due to the difficult mechanism modeling process and are only suitable for offline identification. To solve these problems, this paper proposes a black-box modeling method to identify the voltage and frequency response model of microgrid online. Firstly, the microgrid system is set as a two-input, two-output black-box system and can be modeled only by data sampled at the input and output ports. Therefore, the simplicity of modeling steps can be guaranteed. Meanwhile, the recursive damped least squares method is used to realize the online model identification of the microgrid system, so that the model parameters can be adjusted with the change of the microgrid operating structure, which makes the model more adaptable. The paper analyzes the black-box modeling process of the microgrid system in detail, and the microgrid platform, including 100 kW rated power inverters, is employed to validate the analysis and experimental results.

Keywords: microgrid; recursive damped least squares; black-box modeling; online identification

1. Introduction

Since the microgrid technology can realize flexible and efficient applications of distributed power generations, more and more attention has been paid to its research [1,2]. However, as the scale of microgrid increases, its large-signal behavior becomes complex at the system level analysis. To deal with these problems, microgrid dynamic modeling is one of the most effective methods. At present, microgrid modeling methods mainly include two types: mechanism modeling and identification modeling [3]. The mechanism modeling is carried out on the premise of mastering microgrid control method and structure, and therefore the model has high precision accordingly. Using the constructed model, the small/large signal stability analysis can be performed and the high frequency response characteristics of microgrid system can be studied as well [4,5]. However, with the widespread use of commercial converters, the detailed information relating to converters' topology, parameters, and control methods is hardly to be obtained due to the commercial confidentiality and other reasons, which makes the mechanism modeling method extremely difficult to carry out. Compared with the mechanism modeling, the identification modeling method only needs to sample the input and output data to complete the identification modeling process, instead of mastering the detailed information of microgrid structure and control method. The identified model neglects to some extent the high frequency response characteristics, yet it is simple and convenient to use, thus increasing attention has been paid to it.

Identification modeling method was initially applied in the grid-connected photovoltaic (PV) distributed generation (DG) system modeling [6], and gradually extended to distributed power supplies and distributed power controllers [7–9]. For example, References [10,11] studied identification modeling method of DC/DC converter, and nonlinear models were selected to describe converters' output behaviors. In Reference [12], the black-box modeling process for a single three-phase inverter was proposed, and the dynamic parameters of the system were obtained through the step response. Moreover, the model also took the effects of cross-coupling into account, which further improved the accuracy of the model. In Reference [13], the parameters of the switched reluctance generator were identified, and the influence of nonlinear factors was considered as well, such as PI regulator and clamping functions. Currently, most of the research on microgrid identification modeling is aimed at the certain single converter integrated into the microgrid, such as DC/DC converter, photovoltaic inverter, and so on. By contrast, there are few studies on black-box identification modeling of microgrid at system level. Even for the identification modeling of the whole microgrid, the first step is to build non-linear models of the microgrid components, and then combine them into the microgrid model which is complex in model structure. For instance, Reference [14] described the relationships between the input and output voltage current of each DG by multi-segment nonlinear functions and then combined them together to build up the microgrid model. However, as the scale of the microgrid increases, the adaptability of this method is greatly limited. Reference [15] adopted the non-linear autoregressive method to construct the DGs integrated into the microgrid model by sampling the voltage data at the input port and the fault current at the output port, with up to 11 parameters to be identified.

On the other hand, most identification modeling methods are suitable for fixed microgrid structure, and off-line identification strategies can be employed to model microgrid system [16]. Although under some circumstances (such as similar power supply characteristics, operation mode is basically fixed, etc.), off-line identification has certain applicability, but for the common microgrid structure, there is a random change in power output and load switching, which makes the applicability of this method greatly being challenged. If the online identification method is adopted in identifying microgrid model, the real time update of the model parameters can make up for this deficiency. Nowadays, on-line identification methods are usually adopted in asynchronous motor parameters' identification [17,18], while application of these methods being adopted in microgrid system modeling has not yet been found in literatures.

Therefore, this paper proposes a novel black-box modeling method of microgrid system from a brand-new perspective. The model can be identified only by the voltage frequency data sampled at the point of common coupling (PCC) and the total active and reactive power references data obtained at the input port. The main features are as follows:

1. This method has wide adaptability and is not specific to a particular structure of microgrid. When the structure of microgrid changes randomly, this method is still applicable.
2. The internal structure of the microgrid system and the types of loads are not considered at all. The method only pays attention to the response characteristics observed at PCC to ensure that the model structure is simple and easy to be applied.
3. The proposed method is not confined to the analysis of relationship between power references and the voltage frequency, it is a general idea that can be applied in analyzing other port data of microgrid.
4. Meanwhile, this method is also applicable to all kinds of microgrid systems, such as AC microgrids, DC microgrids, AC/DC hybrid microgrids, and microgrids with various types of electrical equipment and loads (like rotating generators based DG, inverter connected distributed energy resource (DER), controllable and uncontrollable loads, etc.).

2. Topology and Modeling Requirements of Microgrid

2.1. Typical Topology of Microgrid System

The typical microgrid topology is shown in Figure 1. The system commonly comprises DGs such as photovoltaic and wind power generations, as well as large-capacity energy storage devices that maintain the stability of the microgrid voltage and frequency. These devices are connected to the AC bus via the inverters. Some active and reactive loads are also included in the microgrid system. Equipment with communication functions such as specific DGs and smart switches can be connected to the microgrid central controller (MGCC) which can also send control instructions to DGs and smart switches in turn. When the smart switch at PCC is disconnected, the whole microgrid system operates in islanded mode. At this time, the microgrid can adopt a peer-to-peer control structure, and then use the control strategy such as droop control [19] and virtual synchronous machine control [20] to form the voltage and frequency of the microgrid.

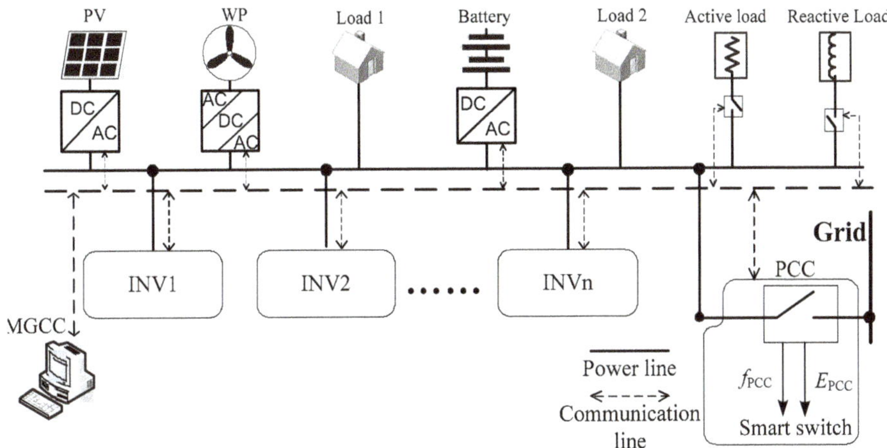

Figure 1. Topological Structure of Microgrid System.

2.2. Modeling Requirements

When studying the coordinated control of the microgrid system, usually the first step is to model and analyze the DGs included in the microgrid system, which is an important part before the overall modeling of microgrid. A rule of thumb is to use the mechanism modeling method to equivalent multiple energy storage inverters of the same type into one inverter so that the microgrid model can be greatly simplified. However, in most microgrid systems, besides the energy storage devices, there are various types of distributed power devices, which affect the system characteristics as well. These factors cannot be ignored, which lead to the model being too complex. Without the information of the internal structure of the microgrid system, the microgrid model can be equivalent to a black-box model and identification modeling method can be adopted to model the microgrid system.

According to the black-box model of microgrid as shown in Figure 2, the parameters optimization of secondary voltage and frequency controller can be carried out with less complexity. The black-box model of microgrid is considered as part of the open-loop transfer function of the overall control structure, and the corresponding root locus can be obtained according to the open-loop transfer function. Hence, the performance of microgrid system can be analyzed and the parameters of PI regulators can be optimized as well.

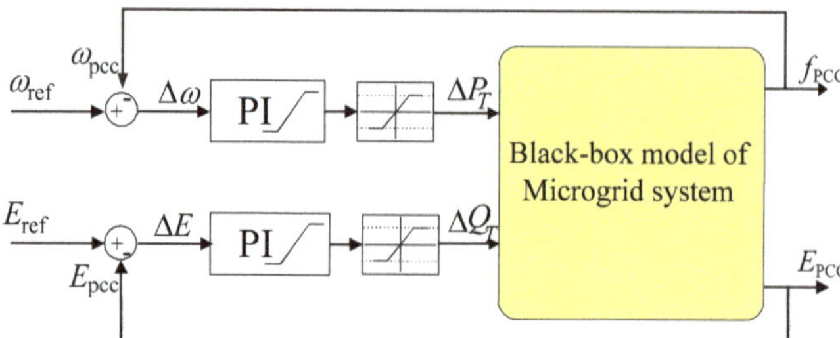

Figure 2. Microgrid Black-Box Model for Secondary Frequency and Voltage Regulation Control Structure.

3. Identification of Black-Box Model

3.1. The Structure of Identification Model

According to the modeling requirements discussed above, the equivalent structure of microgrid model can be obtained as shown in Figure 3. Instead of analyzing the internal structure of the microgrid, the black-box model can be directly constructed by sampling the active and reactive power references data at the input ports and the frequency and voltage amplitude data at the output ports.

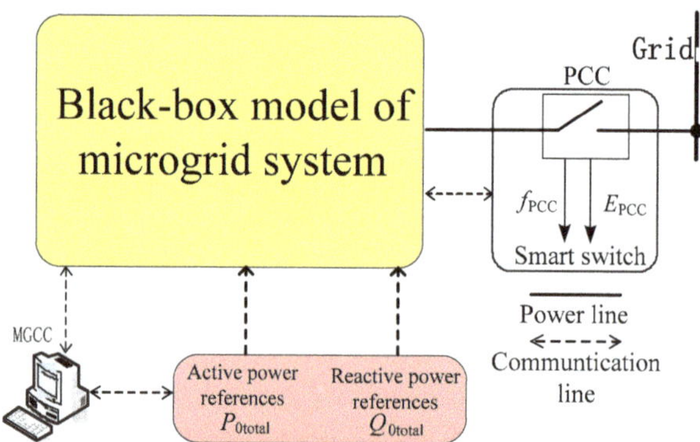

Figure 3. Black-box model structure of microgrid.

In theory, if the line impedance and the output impedance of the inverters included in the microgrid system are purely inductive, the relationship between the frequency and the active power, as well as the relationship between the voltage and the reactive power are completely decoupled. However, in practical microgrid systems, the line impedance and the output impedance are not purely inductive, so the coupling term should be considered in the black-box model. Therefore, there are 4 parameters to be identified in the black-box model as shown in Equation (1):

$$\begin{pmatrix} f_{\text{PCC}} \\ E_{\text{PCC}} \end{pmatrix} = \begin{pmatrix} G_{fP}(s) & G_{fQ}(s) \\ G_{EP}(s) & G_{EQ}(s) \end{pmatrix} \begin{pmatrix} P_{\text{0total}} \\ Q_{\text{0total}} \end{pmatrix} \tag{1}$$

where P_{0total} and Q_{0total} refer to the total active and reactive power references in the microgrid respectively, and according to the power droop coefficients of each DG, MGCC assigns the total power reference values P_{0total} and Q_{0total} to each DG according to Equation (2):

$$\begin{cases} m_1P_1 = m_2P_2 = m_3P_3 = \cdots = m_iP_i \\ n_1Q_1 = n_2Q_2 = n_3Q_3 = \cdots = n_iQ_i \\ P_{0\text{total}} = P_1 + P_2 + P_3 + \cdots P_i \\ Q_{0\text{total}} = Q_1 + Q_2 + Q_3 + \cdots Q_i \end{cases} \tag{2}$$

where m_i and n_i refer to the active and reactive power droop coefficients of ith inverter, P_i and Q_i refer to the active and reactive power references of ith inverter, and f_{PCC} and E_{PCC} are frequency and voltage amplitudes of PCC. G_{fP}(s) and G_{EP}(s) are transfer functions of the active power reference with respect to the frequency and the voltage amplitude of PCC respectively, G_{fQ}(s) and G_{EQ}(s) are the transfer functions of the reactive power reference with respect to the frequency and the voltage amplitude of PCC respectively. Among them, G_{EP}(s) and G_{fQ}(s) are added to the model specifically in order to analyze the cross-coupling effect.

According to Equation (1), the microgrid system is regarded as a two-input and two-output black-box model. Without involving the characteristics of a single DG, the black-box identification model of the whole microgrid system can be obtained as shown in Figure 4.

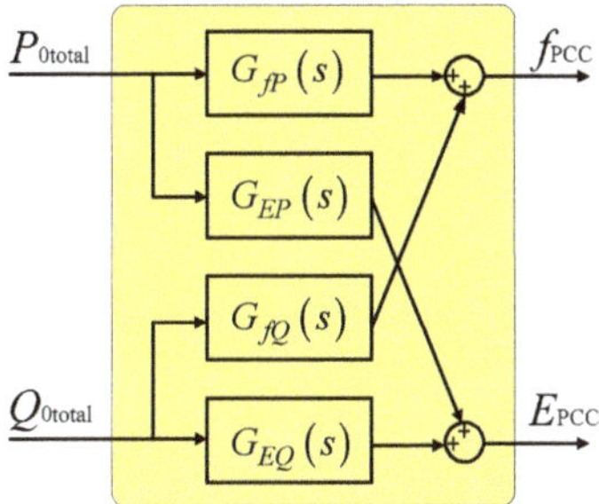

Figure 4. Equivalent black-box model of voltage frequency response of microgrid.

By changing the total active and reactive power references issued by MGCC and sampling the voltage and frequency response data from the smart switch, the voltage and frequency response model of microgrid can be constructed by on-line identification method.

3.2. Identification Experiments Design

Since the step experiment process is easy to realize and has good identification performance, the step response experiment is adopted to identify the transfer function of the model. In order to simplify the identification process of the transfer functions mentioned above, this work divides modeled transfer functions into two groups. As shown in Figure 5, one group carries out the step experiment by performing a step on the active power reference value, the other group carries out the step experiment by performing a step on the reactive power reference value.

3.2.1. Active Power Reference Step

The active power reference step experiment is applied to identify G_{fP}(s) and G_{EP}(s). The $Q_{0\text{total}}$ is set to 0, while the total active power reference is regarded as the input of the step experiment by performing a step on $P_{0\text{total}}$. At the same time, the f_{PCC1} and E_{PCC1} are sampled at PCC as the output of the step experiment.

$$\begin{aligned} G_{fP}(s) &= \frac{f_{\text{PCC}}}{P_{0\text{total}}}\Big|_{Q_{0\text{total}}=0} \\ G_{EP}(s) &= \frac{E_{\text{PCC}}}{P_{0\text{total}}}\Big|_{Q_{0\text{total}}=0} \end{aligned} \tag{3}$$

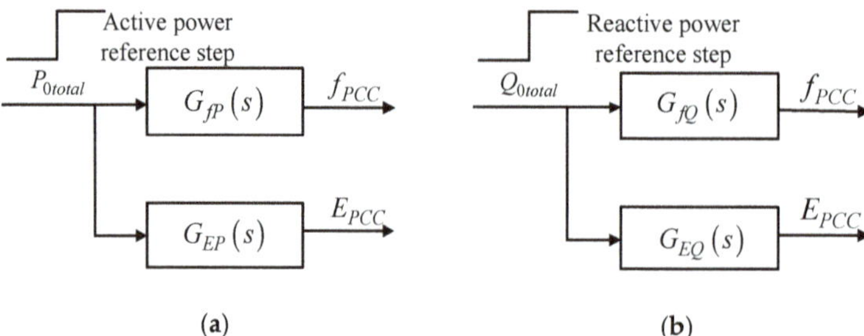

Figure 5. Input power references step experiments (**a**) Active power references step; (**b**) Reactive power references step.

3.2.2. Reactive Power Reference Step

The reactive power reference step experiment is applied to identify $G_{fQ}(s)$ and $G_{EQ}(s)$. The $P_{0\text{total}}$ is set to 0, while the total reactive power reference is regarded as the input of the step experiment by performing a step on $Q_{0\text{total}}$. Meanwhile, the f_{PCC2} and E_{PCC2} are sampled at PCC as the output of the step experiment.

$$\begin{array}{l} G_{fQ}(s) = \frac{f_{\text{PCC}}}{Q_{0\text{total}}}\Big|_{P_{0\text{total}}=0} \\ G_{EQ}(s) = \frac{E_{\text{PCC}}}{Q_{0\text{total}}}\Big|_{P_{0\text{total}}=0} \end{array} \tag{4}$$

3.3. *Identification Strategy and Process*

By sampling the input and output data of the microgrid system, the transfer function expressions of the microgrid model shown in Figure 4 can be obtained through an identification algorithm. The existing identification algorithms usually use the recursive least squares method to identify the system on-line and keep updating the parameters in real time. However, as the covariance matrix decreases in the recursive process, the parameters are prone to explode [21]. In order to enhance the stability of the recursive process, a damping term can be added to suppress the parameter explosion. Therefore, the recursive damped least squares method is used in this paper as the identification algorithm.

Moreover, in the classical control method for discrete systems, the bilinear transformation can ensure that the continuous system and the discrete system have the same stability and the same steady-state gain [22]. Therefore, the bilinear transformation is chosen to discretize the transfer functions shown in Equations (2) and (3), and the recursive damped least squares method is used for the parameter identification.

This paper takes $G_{fP}(s)$ as an example and analyzes the identification process in detail, other transfer functions $G_{EP}(s)$, $G_{fQ}(s)$ and $G_{EQ}(s)$ can adopt the same process.

3.3.1. Data Preprocessing

The first task is to preprocess the sampled data $P_{0\text{total}}$ and E_{PCC1}. It mainly includes two steps:

1. Offset removal: The black-box model of the microgrid system includes the steady-state component and the dynamic component. But the parameters to be identified only involve the system dynamics of the system model, thus it is necessary to remove the steady-state components, i.e., the steady-state values of the input and output signals (before the step is done) respectively.
2. Measurement prefiltering: Since the sampling data contains some ripple and noise caused by the switching frequency, prefiltering input and output data should be considered before the recursive process to avoid the identification accuracy being affected.

3.3.2. Model Order and Initial Value Selection

Secondly, the number of coefficients of each polynomial (transfer function order) and the initial values of parameters should be determined. The first n data to be identified (in order to reduce the computation load, the selected data should not be too much) can be selected, and the least squares algorithm can be used to test a certain transfer function order iteratively. By comparing the fitness between the model output and the measured output in different orders, the order of the transfer function $G_{fP}(s)$ can be determined and remains constant in the process of recursion. The fitting performance can be calculated as follows:

$$f_{\text{Best fit}} = 100\% \times \left(1 - \frac{\sqrt{\sum\limits_{k=1}^{N} (y(k) - \overline{y})^2}}{\sqrt{\sum\limits_{k=1}^{N} (y(k) - \hat{y}(k))^2}} \right) \overline{y} = \frac{1}{N} \sum_{K=1}^{N} y(k) \tag{5}$$

where $y(k)$ is the measured output of microgrid, $\hat{y}$(k) is the model output, $\overline{y}$ is the average value of the measured output. The higher the $f_{\text{Best fit}}$, the better the model can reproduce the output characteristics of the microgrid port. Considering that the high model order affects the computing speed significantly, so as long as the fitness meets the requirements, choosing a lower transfer function order is recommended. Then, the order of the numerator of $G_{fP}(s)$ is set to 1, while that of the denominator is 2, the corresponding expression of $G_{fP}(s)$ can be obtained as

$$G_{fP}(s) = \frac{f_{PCC}(s)}{P_{0total}(s)} = \frac{b_1 s + b_2}{s^2 + a_1 s + a_2} \tag{6}$$

The transfer function can be discretized by the bilinear transformation, as shown in Equation (7):

$$G_{fP} = G_{fP}(s)\Big|_{(s=\frac{Tz-1}{2z+1})} = \frac{B_1 + B_2 z^{-1} + B_3 z^{-2}}{1 + A_1 z^{-1} + A_2 z^{-2}} \tag{7}$$

where

$$\begin{cases} A_1 = \left(-T^2/2 + 2a_2 - a_1 T\right)/\Delta \\ A_2 = \left(T^2/4 + a_1 T/2 + a_2\right)/\Delta \\ B_1 = (b_2 + b_1 T/2)/\Delta \\ B_2 = (2b_2 - b_1 T)/\Delta \\ B_3 = (b_2 + b_1 T/2)/\Delta \\ \Delta = T^2/4 + a_1 T/2 \end{cases} \tag{8}$$

3.3.3. Online Recursive Algorithm

Then the corresponding difference equation of Equation (7) is:

$$f_{PCC}(k) = -A_1 f_{PCC}(k-1) - A_2 f_{PCC}(k-2) + B_1 P_{0total}(k) + B_2 P_{0total}(k-1) + B_3 P_{0total}(k-2) \tag{9}$$

where f_{PCC} (k) and P_{0total} (k) are discrete values of f_{PCC} and P_{0total} at time k.

Let k = 3, 4,..., m, m is the total number of samples during the time of recursive computation. Equation (9) can be rewritten in the form of a matrix:

$$\boldsymbol{F}_m = \boldsymbol{H}_m \boldsymbol{\theta} \tag{10}$$

where $\boldsymbol{F}_m = [f_{PCC}(3), f_{PCC}(4), \cdots, f_{PCC}(m)]^{\mathrm{T}}$, $\boldsymbol{H}_m = [\varphi_3, \varphi_4, \cdots, \varphi_m]^{\mathrm{T}}$, $\theta = [A_1, A_2, B_1, B_2, B_3]^{\mathrm{T}}$, $\boldsymbol{\varphi}_m = [f_{PCC}(m-1), f_{PCC}(m-2), P_{\text{total}}(m), P_{\text{total}}(m-1), P_{total}(m-2)]$.

The idea of the recursive damped least squares method is to revise the estimated value $\widehat{\theta}$ recursively so that the minimum sum of squared errors between F_m and $\widehat{F}_m = H_m\widehat{\theta}_m$ can be obtained:

$$J(\widehat{\theta}) = (F_m - H_m\widehat{\theta}_m)^{\mathrm{T}}(F_m - H_m\widehat{\theta}_m) = \min \tag{11}$$

where $J(\widehat{\theta})$ is the objective function. As discussed above, adding damping to the algorithm can suppress parameter explosion, so a damping term is added on the basis of equation (11):

$$J(\widehat{\theta}) = (F_m - H_m\widehat{\theta}_m)^{\mathrm{T}}(F_m - H_m\widehat{\theta}_m) + \mu\left[\widehat{\theta}_m - \widehat{\theta}_{m-1}\right]^{\mathrm{T}}\left[\widehat{\theta}_m - \widehat{\theta}_{m-1}\right]^{\mathrm{T}} \tag{12}$$

where μ is the damping factor, which is related to the linearity of the identification model.

This paper adopts the adaptive damping factor method to adjust the value of μ. Firstly, the matrix $H_nH_n^T$ is employed to determine the initial value of μ:

$$\mu_0 = 0.01 \times \mathrm{mean}(H_nH_n^T) \tag{13}$$

where the function mean is to compute the average value of the matrix. The linearity of the microgrid black-box model can be quantitatively evaluated by the ratio of the reduction and the change of the objective function J in the iterative process [22]:

$$\lambda = \frac{J(\widehat{\theta}_m) - J(\widehat{\theta}_{m-1})}{\widetilde{J}(\widehat{\theta}_m) - J(\widehat{\theta}_{m-1})} \tag{14}$$

where $J(\widehat{\theta}_m)$ is the calculated value of the objective function J at time m, $\widetilde{J}(\widehat{\theta}_m)$ is the second-order Taylor series expansion of the objective function J at time m:

$$\widetilde{J}(\widehat{\theta}_m) = J(\widehat{\theta}_n) + J'(\widehat{\theta}_n)(\widehat{\theta}_m - \widehat{\theta}_n) + \frac{J''(\widehat{\theta}_n)}{2}(\widehat{\theta}_m - \widehat{\theta}_n)^2 \tag{15}$$

where J' is the first derivative of J, J'' is the second derivative of J.

Therefore, the damping factor can be corrected using the value calculated in Equation (14):

1. Under the condition of $\lambda < 0.3$, it shows that the objective function J exhibits a linearity reduction trend during the recursive process, μ is needed to increase in the subsequent recursive process to ensure better frequency and voltage identification
2. Under the condition of $0.3 < \lambda < 0.7$, it indicates that the objective function J has a small change in linearity during the recursive process, therefore the damping factor is close to the best value and there is no need to change it.
3. Under the condition of $\lambda > 0.7$, a smaller damping factor μ should be chosen.

In order to achieve the adaptive correction of damping factor during the identification process, Proportional coefficient η is employed, whose value is generally between 2 and 10, as shown below:

$$\eta = \frac{J(\widehat{\theta}_m) - J(\widehat{\theta}_{m-1})}{(f_{\mathrm{PCC}}(m) - \varphi(m)\cdot x_k)\cdot x_k} + 2 \tag{16}$$

when μ is needed to be increased, μ can be taken as

$$\mu = \mu_0 \cdot \eta \tag{17}$$

on the contrary, when μ is needed to be decreased, μ can be taken as

$$\mu = \mu_0/\eta \tag{18}$$

According to the extremum principle, the minimum value of J can be obtained by taking the differentiation to the right part of Equation (12). Through combing Equations (10)–(12), the recursive equation based on recursive damped least square method is derived as below:

$$\begin{cases} \boldsymbol{P}_{m+1} = \left[\mu \boldsymbol{I} + \boldsymbol{H}_{m-1}^{\mathrm{T}} \boldsymbol{H}_{m-1} + \varphi_m^{\mathrm{T}} \varphi_m\right]^{-1} \\ \hat{\boldsymbol{\theta}}_{m+1} = \hat{\boldsymbol{\theta}}_m + \mu \boldsymbol{P}_{m+1}\left[\hat{\boldsymbol{\theta}}_m - \hat{\boldsymbol{\theta}}_{m-1}\right] + \boldsymbol{P}_{m+1} \boldsymbol{h}_{m+1}^{\mathrm{T}}\left[f_{\mathrm{PCC}}(m) - \varphi_{m+1} \hat{\boldsymbol{\theta}}_m\right] \end{cases} \tag{19}$$

Ever since a set of observation data φ_m is added, $\boldsymbol{P}_{\mathrm{m}}$ and $\hat{\theta}_m$ are revised once, therefore new estimation values of parameters are derived and on-line identification of parameters are realized. The identification procedure is shown in Figure 6. Where ε is the allowable range of precision error, N is the total number of samples at the current moment. When the recursive algorithm compute to the sampling moment or the parameters estimation does not change, the recursive process is terminated until the new sampling data arrives. After obtaining the parameters of the transfer function $G_{fP}(z)$, the bilinear inverse transformation can be carried out to convert the transfer function from discrete domain to continuous domain.

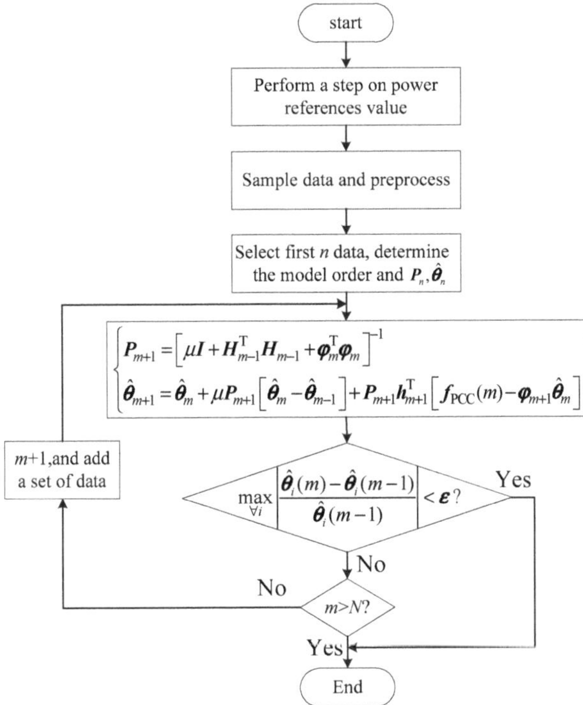

Figure 6. Flowchart of identification procedure.

4. Experiments and Model Validation

In this paper, a series of model identification experiments are carried out on a microgrid test platform to verify the accuracy of the proposed modeling method. The controllers of the two inverters integrated into the microgrid use MCU (TMS320F28335, Texas Instruments, Inc, Dallas, TX, USA), the pulse width modulation (PWM) carrier frequency of each controller is set to 6 kHz, and the integrated three-phase resistive load is configured to 30 kW. Both inverters operate in *P-f* droop control mode. Active and reactive power droop coefficients are chosen as 4×10^{-6} and 4×10^{-4} respectively. Under the normal operation of the experimental platform, the input and output data are sampled by the Yokogawa recorder. All devices are connected to the AC Bus and communication lines, the structure of the microgrid test platform is shown in Figure 7.

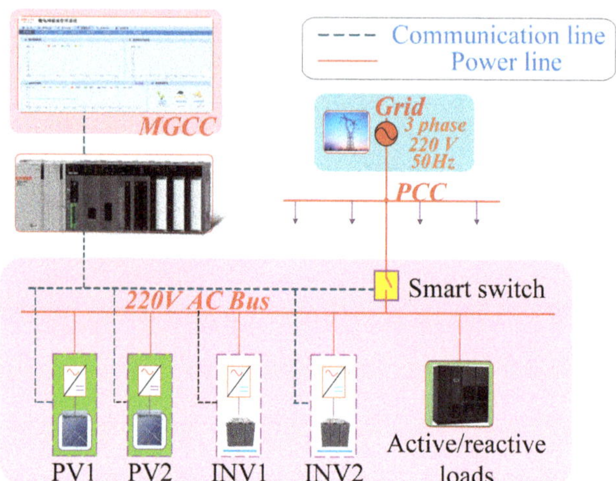

Figure 7. The structure of microgrid test platform.

4.1. Identification Experiments of $G_{fP}(s)$ and $G_{EP}(s)$

To carry out the identification experiments of $G_{fP}(s)$ and $G_{EP}(s)$, parallel inverters and the three-phase active load are integrated into the microgrid. After the system reaches a steady state, the active and reactive power references are set to 0 by MGCC at first. Then the active power reference is changed to 80 kW at 9 s, and set to 0 again at 15.7 s. During this experiment, the reactive power reference remains at 0. Figure 8 shows the identification data P_{0total}, E_{PCC1}, and f_{PCC1} sampled by the waveform recorder at PCC.

According to the identification process shown in Figure 6, the sampled data P_{0total} and f_{PCC1} can be used to identify the model transfer function $G_{fP}(s)$.

$$G_{fP}(s) = \frac{4.229 \cdot 10^{-8}s^2 - 0.001s + 5.0184}{s^2 + 945.97s + 2.5172 \cdot 10^5} \tag{20}$$

By applying the same input on both the model and the microgrid system, the fitting performance can be obtained as shown in Figure 9. Since only the dynamic component is analyzed, the steady-state component contained in each measurement has to be removed. As it can be seen, the model output matches the microgrid output properly, and $f_{\text{Best fit}}$ reaches 95.36.

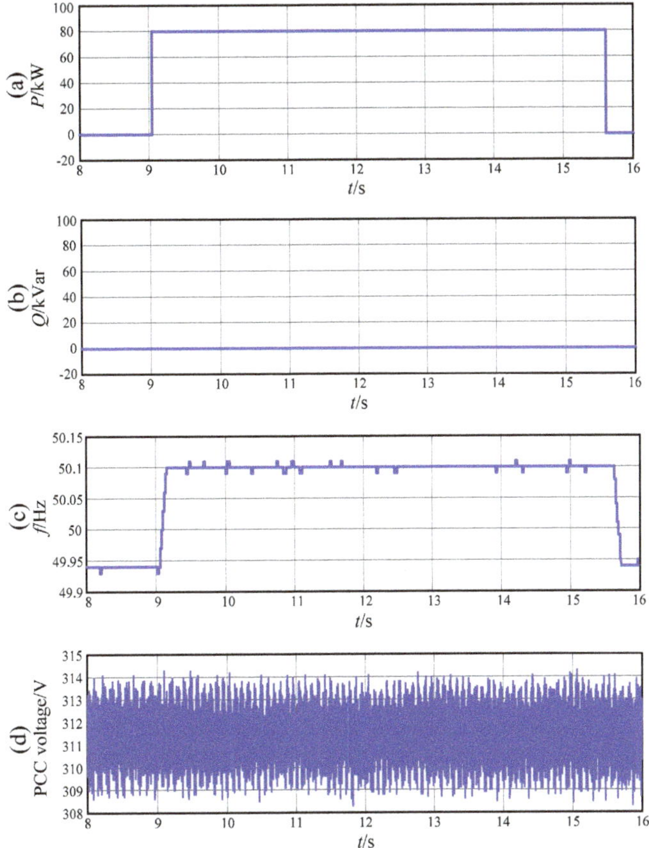

Figure 8. Active power step sample data for identification (**a**) Active power references; (**b**) Reactive power references; (**c**) PCC frequency; (**d**) PCC voltage amplitude.

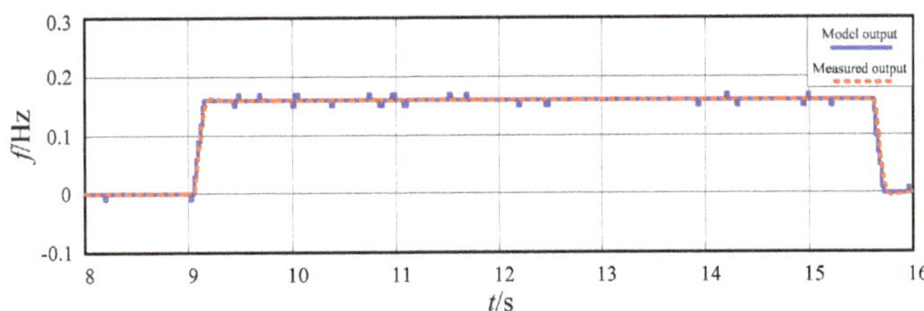

Figure 9. Fitting curves of the transfer function model obtained from the active power step experiment.

Similarly, using the sampled data P_{0total} and E_{PCC1}, the transfer function $G_{EP}(s)$ can be obtained:

$$G_{EP}(s) = \frac{0.06107s^2 + 64.24s + 152.9}{s^3 + 1249s^2 + 3.95 \cdot 10^5 s + 4.933 \cdot 10^8} \tag{21}$$

4.2. Identification Experiments of $G_{fQ}(s)$ and $G_{EQ}(s)$

To carry out the identification experiments of $G_{fQ}(s)$ and $G_{EQ}(s)$, parallel inverters and the three-phase reactive load are integrated into the microgrid. After the system reaches a steady state, the active and reactive power references are set to 0 by MGCC at first. Then the reactive power reference is changed to 80 kVar at 9 s, and set to 0 again after 5.6 s. During this experiment, the active power reference remains at 0. Figure 10 shows the identification data $Q_{0\text{total}}$, E_{PCC2}, and f_{PCC2} sampled by the waveform recorder at PCC.

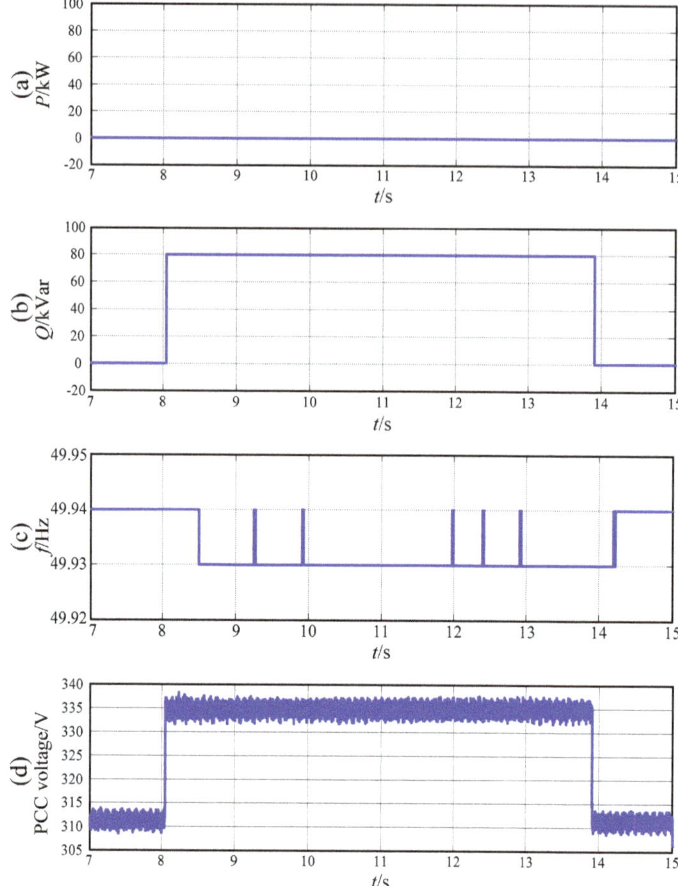

Figure 10. Reactive power step sampling data for identification (**a**) Active power references; (**b**) Reactive power references; (**c**) PCC frequency; (**d**) PCC voltage amplitude.

According to the identification process shown in Figure 6, the sampled data $Q_{0\text{total}}$ and E_{PCC2} can be used to identify the model transfer function $G_{EQ}(s)$.

$$G_{EQ}(s) = \frac{1.0129 \cdot 10^{-5}s^2 - 1.288s + 7826.65}{s^2 + 2760.77s + 2.545 \cdot 10^6} \tag{22}$$

By applying the same excitation on both the model and the microgrid system, the fitting performance can be obtained as shown in Figure 11. Clearly, the model output also matches the microgrid output, and $f_{\text{Best fit}}$ reaches 90.72.

Similarly, using the sampled data $Q_{0\text{total}}$ and f_{PCC2}, the transfer function $G_{fQ}(s)$ can be obtained as follows:

$$G_{fQ}(s) = \frac{8.085 \cdot 10^{-6}s^2 + 2.219 \cdot 10^{-4}s + 6.091 \cdot 10^{-4}}{s^3 + 8.47s^2 + 710.6s + 5187} \tag{23}$$

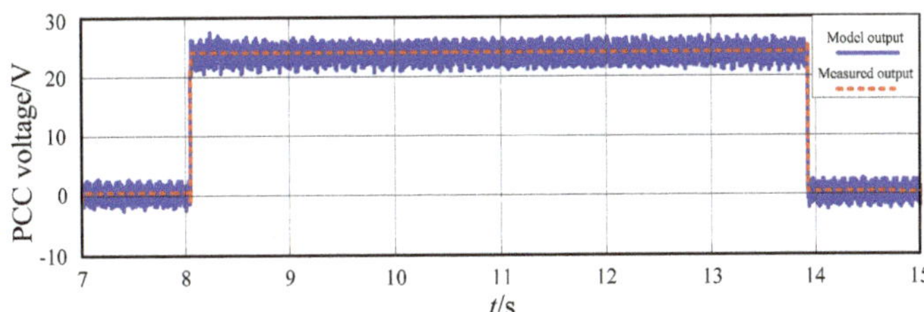

Figure 11. Fitting curves of the transfer function model obtained from the reactive power step experiment.

4.3. Recursive Model Validation

According to the power references step experiments, the voltage frequency response model is obtained under the certain microgrid structure. Furthermore, in order to verify the adaptability of the recursive damped least squares method applied in the model identification, that is, the online identification effect during the period of microgrid structure changing, several adjustments have been made: active power droop coefficients of two inverters are changed to 4×10^{-6} and 8×10^{-6} respectively; reactive power droop coefficients are changed to 2×10^{-4} and 4×10^{-4} respectively; the PV inverter which is supplied by Chroma solar PV array simulators is added to the microgrid structure, and the maximum output power of the PV inverter is set to 4 kW. During the power references step experiment, the number of inverters connected to the microgrid are changed, the active and reactive loads are set to different parameters, as listed in Table 1.

Table 1. Experimental parameters.

Time	Total Power References	Active Load	Reactive Load
t_1	0/0	0	0
t_2	0/0	12 kW	20 kVar
t_3	60 kW/40 kVar	12 kW	20 kVar
t_4	100 kW/60 kVar	12 kW	20 kVar
t_5	60 kW/40 kVar	12 kW	20 kVar
t_6	60 kW/40 kVar	8 kW	10 kVar
t_7	60 kW/40 kVar (INV1 shutdown)	8 kW	10 kVar
t_8	60 kW/40 kVar (INV1 start up)	8 kW	10 kVar
t_9	0/0	8 kW	10 kVar
t_{10}	0/0	0	0

The input and output data during the experiment are recorded by the waveform recorder, as shown in Figure 12.

Meanwhile, the same experimental processes as shown in Table 1 are carried out on a corresponding simulation model under the MATLAB/Simulink environment (2014a, The MathWorks, Inc, Natick, MA, USA), which is constructed according to the black-box equivalent model shown in Figure 4. Initial values of $\boldsymbol{P}$ and $\hat{\theta}$ can be figured out according to the transfer function parameters obtained in Sections 3.2 and 4.1. The frequency and voltage output of the simulation model can be obtained by using the recursive damped least squares method for online identification.

The experimental and simulation results are compared with each other, as shown in Figure 13. It can be seen that the simulation model is able to reproduce the system responses of the microgrid system during the experiment, especially when the frequency and voltage responses of the microgrid have step changes, the output of the simulation model tracks the changes accurately. Hence, the identification algorithm implemented in the simulation model is validated.

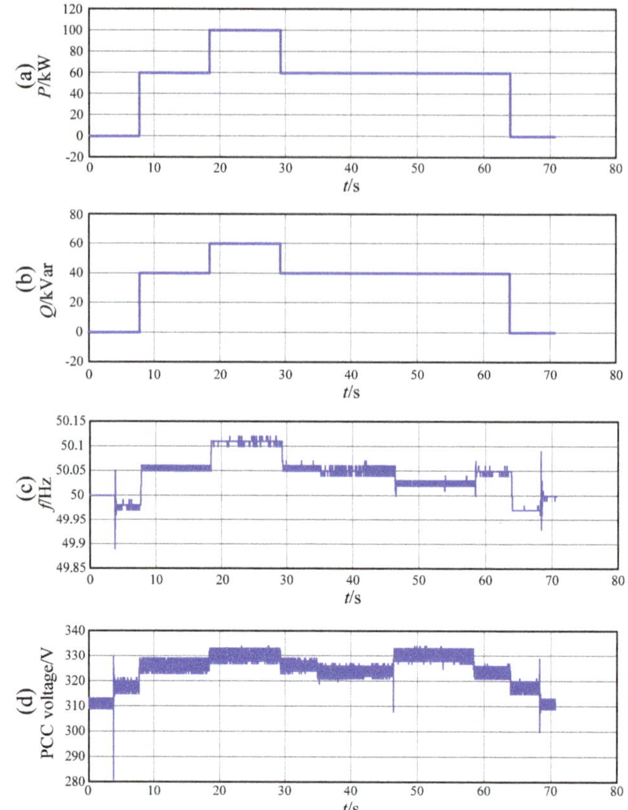

Figure 12. Data waveforms sampled at input and output ports of the microgrid (**a**) Active power references; (**b**) Reactive power references; (**c**) PCC frequency; (**d**) PCC voltage amplitude.

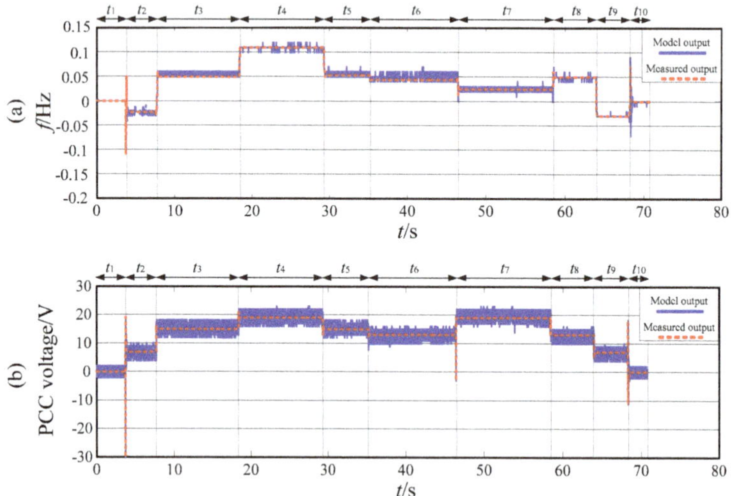

Figure 13. Identification model and measured output curves (**a**) PCC frequency; (**b**) PCC voltage amplitude.

5. Conclusions

In this paper, the black-box modeling technique of voltage frequency response model of microgrid system has been studied. In terms of the parameter identification algorithm of the recursive damped least squares, the model can be identified and is suitable for performing simulations of microgrid system on system level, especially when the internal information of a microgrid cannot be accessed. The procedure of constructing the black-box model of microgrid is analyzed emphatically. Compared with the traditional method of mechanism modeling, this method greatly simplifies the modeling steps and provides a general idea for microgrid system modeling.

The output data obtained from the simulation model has been compared with the measured output under the conditions of changing the microgrid structure. Depending on the recursive algorithm, the identification model can reproduce the frequency and voltage characteristics of the microgrid port accurately in all cases, which indicates that the black-box identification modeling method has a wide range of adaptability to realize on-line identification of the microgrid system.

In future, in order to improve the identification accuracy, the influence of different communication delays on the proposed identification strategy needs to be further studied. In addition to this, how to apply the proposed identification model to a secondary frequency and voltage regulation control system, so as to optimize parameters of secondary voltage and frequency controllers, are also set as the future research directions.

Author Contributions: This paper was a collaborative effort between the authors. Y.S. provided the original idea and reviewed the paper, D.X. and H.X. organized the manuscript and realized MATLAB-based simulation, H.Y. carried out experiment validation, J.S. and N.L. provided professional advice and technical comments.

Funding: This research was funded by Double First Class Project for Independent Innovation and Social Service Capabilities (45000-411104/012), National Key Research and Development Program of China (2017YFB0903503) and the Basic Operating Expenses of Central Scientific Research (PA2018GDQT0021).

Conflicts of Interest: The authors declare no conflict of interest.

References

1. Yu, X.D.; Xu, X.D.; Chen, S.Y.; Wu, J.Z.; Jia, H.J. A brief review to integrated energy system and energy internet. *Trans. Chin. Electrotech. Soc.* **2016**, *31*, 1–13.
2. Hou, X.B.; Wang, L.; Li, Q.M.; Wang, J.; Liu, J.Z.; Qian, X.S. Review of key technologies for high-voltage and high-power transmission in space solar power station. *Trans. China Electrotech. Soc.* **2017**, *15*, 61–73.
3. Zheng, W. Identification Modeling Method and Application of Photovoltaic Grid-Connected Inverter. Ph.D. Thesis, Chongqing University, Chongqing, China, 2014.
4. Wang, X.; Blaabjerg, F.; Chen, Z. Autonomous control of inverter interfaced DG units for harmonic current filtering and resonance damping in an islanded microgrid. *IEEE Trans. Ind. Appl.* **2014**, *50*, 452–461. [CrossRef]
5. Jadhav, G.N.; Changan, D.D. Modeling of inverter for stability analysis of microgrid. In Proceedings of the IEEE 7th Power India International Conference (PIICON), Bikaner, India, 25–27 November 2016; pp. 1–6.
6. Xiong, X.F.; Chen, K.; Zheng, W.; Shen, Z.J.; Shahzad, N.M. Photovoltaic inverter model identification based on least squares method. *Power Syst. Prot. Control* **2012**, *22*, 52–57.
7. Jin, Y.Q.; Ju, P.; Pan, X.P. A stepwise method to identify controller parameters of photovoltaic inverter. *Power Syst. Technol.* **2015**, *39*, 594–600.
8. Zheng, W.; Xiong, X.F. A model identification method for photovoltaic grid-connected inverters based on the Wiener model. *Proc. CSEE* **2013**, *33*, 18–26.
9. Zheng, W.; Xiong, X.F. System identification for NARX model of photovoltaic grid-connected inverter. *Power Syst. Technol.* **2013**, *39*, 2440–2445.
10. Valdivia, V.; Barrado, A.; Lázaro, A.; Sanz, M.; del Moral, D.L.; Raga, C. Black-Box behavioral modeling and identification of DC–DC converters with input current control for fuel cell power conditioning. *IEEE Trans. Ind. Electron.* **2014**, *61*, 1891–1903. [CrossRef]

11. Frances, A.; Asensi, R.; Garcia, O.; Prieto, R.; Uceda, J. How to model a DC microgrid: towards an automated solution. In Proceedings of the IEEE Second International Conference on DC Microgrids (ICDCM), Nuremburg, Germany, 27–29 June 2017.
12. Valdivia, V.; Lazaro, A.; Barrado, A.; Zumel, P.; Fernandez, C.; Sanz, M. Black-box modeling of three-phase voltage source inverters for system-level analysis. *IEEE Trans. Ind. Electron.* **2012**, *59*, 3648–3662. [CrossRef]
13. Valdivia, V.; Todd, R.; Bryan, F.J.; Barrado, A.; Lázaro, A.; Forsyth, A.J. Behavioral modeling of a switched reluctance generator for aircraft power systems. *IEEE Trans. Ind. Electron.* **2014**, *61*, 2690–2699. [CrossRef]
14. Francés, R.; Asensi, R.; García, O.; Prieto, R.; Uceda, J. The performance of polytopic models in smart DC microgrids. In Proceedings of the IEEE Energy Conversion Congress and Exposition (ECCE), Milwaukee, WI, USA, 18–22 September 2016; pp. 1–8.
15. Erlinghagen, P.; Wippenbeck, T.; Schnettler, A. Modelling and sensitivity analyses of equivalent models of low voltage distribution grids with high penetration of DG. In Proceedings of the 50th International Universities Power Engineering Conference (UPEC), Stoke on Trent, UK, 1–4 September 2015; pp. 1–5.
16. Francés, A.; Asensi, R.; García, O.; Uceda, J. A blackbox large signal Lyapunov-based stability analysis method for power converter-based systems. In Proceedings of the IEEE 17th Workshop on Control and Modeling for Power Electronics (COMPEL), Trondheim, Norway, 27–30 June 2016; pp. 1–6.
17. Cao, P.P.; Zhang, X.; Yang, S.Y.; Xie, Z.; Guo, L.L. Online rotor time constant identification of induction motors based on Lyapunov stability theory. *Proc. CSEE* **2016**, *36*, 3947–3955.
18. Yang, S.Y.; Sun, R.; Cao, P.P.; Zhang, X. Double compound manifold sliding mode observer based rotor resistance online updating scheme for induction motor. *Trans. Chin. Electrotech. Soc.* **2018**, *33*, 3596–3606.
19. Chen, J.; Liu, M.A.; Chen, X.; Niu, B.W.; Gong, C.Y. Wireless parallel and circulation current reduction of droop-controlled inverters. *Trans. Chin. Electrotech. Soc.* **2018**, *33*, 1450–1460.
20. Tu, C.M.; Yang, Y.; Lan, Z.; Xiao, F.; Li, Y.T. Secondary frequency regulation strategy in microgrid based on VSG. *Trans. Chin. Electrotech. Soc.* **2018**, *33*, 2186–2195.
21. Zhao, X.; Wang, Z.P.; Jia, H.H. Comparison and analysis of discretization methods for continuous systems. *Ind. Inf. Technol. Educ.* **2015**, *12*, 71–76.
22. Su, Y.C. The Research of Leven-Marquadt-Fletcher Method for Fitting Transient Electromagnetic Data. Ph.D. Thesis, Jilin University, Shenyang, China, 2009.

Article

Sensor-Based Early Activity Recognition Inside Buildings to Support Energy and Comfort Management Systems

Francesca Marcello [1,†], Virginia Pilloni [1,2,*,†] and Daniele Giusto [1,2]

1 Department of Electrical and Electronic Engineering (DIEE), University of Cagliari, 09123 Cagliari, Italy
2 National Telecommunication Inter University Consortium (CNIT), Research Unit of Cagliari, 09123 Cagliari, Italy
* Correspondence: virginia.pilloni@diee.unica.it
† These authors contributed equally to this work.

Received: 20 June 2019; Accepted: 5 July 2019; Published: 9 July 2019

Abstract: Building Energy and Comfort Management (BECM) systems have the potential to considerably reduce costs related to energy consumption and improve the efficiency of resource exploitation, by implementing strategies for resource management and control and policies for Demand-Side Management (DSM). One of the main requirements for such systems is to be able to adapt their management decisions to the users' specific habits and preferences, even when they change over time. This feature is fundamental to prevent users' disaffection and the gradual abandonment of the system. In this paper, a sensor-based system for analysis of user habits and early detection and prediction of user activities is presented. To improve the resulting accuracy, the system incorporates statistics related to other relevant external conditions that have been observed to be correlated (e.g., time of the day). Performance evaluation on a real use case proves that the proposed system enables early recognition of activities after only 10 sensor events with an accuracy of 81%. Furthermore, the correlation between activities can be used to predict the next activity with an accuracy of about 60%.

Keywords: activity recognition; activity detection; activity prediction; smart building; energy and comfort management

1. Introduction

Smart buildings are characterized by the presence of sensors, actuators, and smart devices that give the opportunity to monitor and remotely control key equipment within buildings [1]. This is the concept behind Smart Building Energy and Comfort Management (BECM) systems [2,3]. In such an intelligent scenario, one of the major goals is to provide decision-support tools that support users in making cost-effective decisions in terms of energy consumption [4]. As a matter of fact, domestic electricity usage accounts for about 40% of the global energy consumption and contributes over 30% of total greenhouse gas emissions [5]. Nevertheless, user comfort is crucial when policies of Demand-Side Management (DSM) are put in place [6]. Indeed, a system that optimizes energy consumption without considering user preferences and habits about appliance usage quickly leads to user disaffection from the system and abandonment of it. Currently, most of the literature considers user comfort as a set of hard constraints on appliance usage, which are a priori set considering general statistics [7,8]. This approach neglects the fact that users are likely not only to have different subjective requirements with respect to the others, but they also dynamically change over time.

In this paper, user preferences and habits about appliance usage are inferred by monitoring them using sensors deployed inside the reference buildings. The first phase consists of analyzing the correlation between the information gathered from sensors and users' activities. The approach proposed in this paper is based on the classifier presented by Krishnan and Cook in [9], which recognizes activities based on sequences of sensor events. Nevertheless, as it will be better explained in the following, in order to improve accuracy, the classifier included in the proposed framework is enhanced with other significant information provided by statistics related to the monitored events. Accordingly, a profile specific to the monitored user is created, which enables early recognition of activities after only 10 sensor events with an accuracy of 81%. Furthermore, the correlation between activities can be used to predict the next activity with an accuracy of about 60%.

The main contributions provided by this paper can be summarized as follows:

- an analysis is performed to study the correlation not only between users' activities and sensor events, but also between sensor events and other components that describe the context, and the mutual correlation of activities;
- an activity recognition algorithm is proposed. The correlation between sensor events and components describing the context is used in the classifier to improve the accuracy of the activity recognition algorithm;
- based on the results of the activity recognition algorithm and statistics about mutual correlation of activities, subsequent activities can be predicted;
- a framework that uses activity recognition and prediction as the main component of a BECM system is described.

The remainder of the paper is organized as follows. Section 2 presents past works and the required background. In Section 3 an overview of the considered system model is provided. Section 4 presents in detail the activity recognition algorithm that enables the early detection and prediction of users' activities. Section 5 describes the reference use case considered to test the performance of the system. Finally, in Section 6 a performance analysis is provided, and conclusions and final remarks are drawn in Section 7.

2. Background

2.1. Smart Building Energy and Comfort Management Systems

Smart technologies can be used in all kinds of different buildings (i.e., residential, office, and retail sectors) to improve the comfort and the safety of people in their home, concerning various topic, from healthcare and providing living assistance, to environmental monitoring and ensuring energy saving. Accordingly, BECM systems have the objective of combining power consumption minimization while preserving user comfort [2,10]. This issue has been addressed by researchers from many different perspectives. In [11] a system for intelligent energy management in buildings is proposed. It presents semantic modeling that integrates all the entities that constitute the environment of a Smart Building exploiting Internet of Thing (IoT) paradigm. The IoT-based system integrates different systems and makes use of various types of real-time data from different sources to achieve the common objective of the intelligent management of the building. In [12] a tool that provides effective automation and control of heating/cooling, ventilation/air conditioning and lighting and that uses optimization techniques to minimize energy consumption is proposed. The interactive system presented achieves a significant decrease in the operating cost of A/C system in a tertiary sector building, while maintaining desirable comfort taking into account two different time periods (peak hours and non-peak hours), a number of different zones and the end-user's preferences. The authors in [13] propose a multi-agent control system for integrated buildings and microgrids, which exploits Renewable Energy Sources (RES) efficiently among the agents. In [7], an algorithm for Distributed Energy Resources (DER) management is proposed to shave peak demand and increase energy efficiency in smart home environments. Customer comfort is considered to be a constraint on time preference ranges to run their appliances. A similar approach was presented in [8], where user preferences are expressed also in terms of indoor temperature and lightning. Furthermore, dynamic pricing is considered to optimize DSM. The authors in [2], after a review of control systems for energy management and comfort in buildings, also present the architecture of a multi-agent control system that manages the user's preferences for thermal and lighting comfort, indoor air quality, and energy conservation. The system is based on a master-slave coordination mechanism to perform different tasks. In [14], an algorithm for thermostatically controlled household loads based on price and consumption forecasts of grid energy is presented. The optimization by means of the algorithm proposed in [14] takes into account the trade-off between customer comfort and cost of energy, by setting minimum and maximum boundaries for the thermal comfort. These boundaries are taken as hard constraints for a priori setting but a background on customer comfort profiling lacks. Collotta and Pau propose a BECM system in [15], where the management is based on a Fuzzy Logic Controller (FLC) which adaptively adjusts appliance starting times based on users' feedback. The issue of scheduling appliances according to user preferences was also addressed by [4], where Quality of Experience (QoE) is measured as a function of the interval between the preferred and proposed appliance starting time for switching controlled loads (e.g., washing machines and clothes dryers), and as a function of the interval between the preferred and proposed temperature for thermostatically controlled loads (e.g., Heating, Ventilation and Air Conditioning (HVAC) and water heaters). The system takes into account both dynamic pricing and RES production.

It is evident that user preferences and habits severely affect results of BECM systems. Indeed, BECM systems that only consider energy cost minimization may switch appliances on/off too early or too late. This is the case, for example, of a dishwasher that does not finish its cycle before dinnertime, or an HVAC that is switched off earlier than what the user considers thermal comfort. For this reason, in recent years researchers have started to observe users' behavior, in order to infer their habits and preferences.

2.2. *Activity Recognition in Smart Buildings*

As discussed in the previous subsection, BECM systems should be able to discover and predict users' habits and course of actions. The monitoring of activities of people in their home can be done by analyzing data that can be gathered with different technologies. Given the rapid development of sensor technology, wireless transmission technology, network communication technology, cloud computing, and smart mobile devices, large amounts of digitized information have been accumulated and the volume of data is growing rapidly with increasingly complex structures and forms [16]. The energy big data provides a new way to analyze and understand individuals' energy consumption behavior, improve energy efficiency, and promote energy conservation. Integration of big data technologies will help make the grid more efficient and it will fundamentally change the way in which regulators, utilities, grid operators, and end-users would interact [17,18].

In many cases, cameras and wearable sensors are used to collect all the information of interest and to understand what someone is doing [19]. These solutions present some problems because people are often not inclined to accept those devices [20]. Some studies are based on the data that are provided by phone accelerometer and gyroscope to understand repetitive body motions (walking, running, sitting) [21]. This solution is not very practical in home scenarios, where residents do not always take their phone with them. To monitor what activities people are performing in their house, non-intrusive sensors are often preferred: typical devices that are installed in the environment are motion sensors, door sensors or temperature and pressure sensors [22,23].

The authors in [24] propose the use of 3D depth sensors to detect user occupancy and profile their habits. The aim of the paper is to use this information to control HVAC and lighting, according to occupants' usual behavior. The relation between occupancy, energy consumption and users' comfort is also investigated in [25], where the Multi-Agent Comfort and Energy Management System is proposed. Along with building devices, MACES also considers occupants as active participants in the building energy reduction strategy and attempts to implement more energy conscious occupant planning. This occupant planning is carried out using multi-objective Markov-Decision Problems (MDPs) to model the uncertainty of agent decisions and interactions.

An interesting approach for energy-consuming activity recognition is proposed in [26], where social media posts are analyzed to automatically extract information and describe energy-consuming activities.

In [27] the authors focus on the activity discovering problem, proposing an approach to build a model under the form of Hidden Markov Model (HMM), from a training database of observed events emitted by binary sensors, without the knowledge of actions really performed during the learning period. The discovering problem is presented also in [28,29], where a Discontinuous Varied-Order Sequential Miner (DVSM) algorithm is used to discover frequent activities that are continuously recorded in a smart environment and combined with a clustering algorithm to find frequent occurrences of activities and cluster familiar patterns together.

In [9], 4 algorithms that can identify the activities while they are being performed are proposed. In this work, activities are recognized even if they are done in an interleaved and concurrent manner. The models used in the proposed algorithms are a Naïve Bayes Classifier, an HMM with a time window, a frequency-based HMM with a sliding window and a frequency-based HMM with a shifting window. These algorithms have been later used in [30] to predict activities with the aim of controlling buildings to reduce energy consumption.

The problem of multi-resident activity recognition based on the use of non-intrusive sensors, along with smartphone-based sensed data, is also addressed in [31]. The approach used in this paper aims to exploit body-worn smartphone sensors to infer person-specific context that can be correlated with activity detected from ambient sensors. Other models for multi-resident activity recognition based on the use of non-intrusive sensors are proposed in [32]. In this paper the authors adopted three different directed graphical models including Poisson HMMs, coupled HMMs, and dynamic Bayesian networks, extended from coupled HMM by adding some vertices, to identify both individual and cooperative activities. A solution for multi-resident environments is given in [33] too, where they perform the tracking of people and recognition of activities by using different binary sensors and RFID to know the identity of the occupants as they enter or leave the environment.

2.3. Background on Activity Recognition

The data collected from sensors inside resident houses are analyzed using data mining and machine learning techniques to build activity models that are used as the basis of behavioral activity recognition. Feature extraction from the sequence of sensor events is a key step to better modeling and then recognizing human activities. In [34] four methods used to extract features for online recognition on streamed data are presented. With their approach, the authors can recognize activities while new sensor events are recorded.

A comparison of classification approaches for activity recognition is provided in [35,36]. As introduced in the previous subsection, with reference to modeling and classification methods researchers have investigated the recognition of resident activities using a variety of mechanisms, such as naïve Bayes classifiers, Markov models, and dynamic Bayes networks.

The various models and algorithms can give different performance results depending on the input data or on the implemented system. Usually, the results are adequately comparable, with one model giving better results in the recognition of some specific activities or using one specific dataset, and others giving better performance recognizing other activities or analyzing other datasets [37]. In multiple cases, in spite of its simple design and simplified assumptions, naïve Bayes classifiers often work much better than expected, especially when a specific group of sensors can easily be identified as characteristic of a certain activity [9].

Starting from the approach presented in [9] and from the analysis of sequences of sensor events for activity recognition solution, in this paper a system that adds statistics information about the context in which these activities are occurring, in order to improve their own recognition, is proposed.

The ultimate goal of the system is to obtain a profile according to users' habits, which allows gaining personalized management of the whole system. Many systems in the literature propose management strategies focused on energy consumption and consider users' comfort, but they tend not to evaluate user activities in every daily aspect in order to learn specific habits and behavior and guarantee ad-hoc solutions.

3. System Model

The reference scenario considered in this paper is that of a BECM system that leverages distributed smart home sensor networks to elaborate user profiles that are later used to manage and control buildings. More specifically, sensors are used to make observations, report events that are detected in the building and that can be associated with users' actions, and learn which sets of detected events can identify specific activities. Therefore, the BECM system can predict users' activities based on their previous monitored actions, and make appropriate management decisions accordingly.

A fundamental step to achieve energy cost savings is the identification of possible causes of energy waste. The BECM system controls all energy consumption components (electric, gas, water) inside the buildings and plans specific actions aimed at reducing waste. A large number of home devices increase power consumption in two aspects: standby power and normal operation power. When high energy demand is detected, the less important loads, such as standby power mode devices, could be disconnected from the electricity grid. Around 10% of the total household power is consumed during the standby power mode [38]. The reduction of standby power is greatly necessary to reduce the electricity cost at home. Thanks to IoT networks, appliances such as washing machines and dishwashers can gather information about different energy prices related to different times of the day, thereby the system can automatically program them for the most convenient times. Air conditioning and heating systems should be monitored to guarantee desirable temperatures in all environments and to avoid unnecessary use. The BECM system records the inhabitants' preferences adapting itself to their way of life.

A system of this type must also be able to know the users' habits, in order to make coherent scheduling in the management of equipment and appliances, and because their activities and behavior have a considerable impact on energy consumption.

The whole system is therefore divided into several modules, each of them with specific features depending on the different tasks pertaining to them. Figure 1 shows the overall architecture of the system and how the modules interact with each other.

The "Acquisition Module" is the one that gets raw data from the sensors. The data are stored in a structured database with information about the sensors and their value, as well as the date and time of the relative information. These data are then processed and manipulated by the "Activity Recognition Module". This module can be organized into two separate tasks. At first, the data are used to understand users' behavior and to make models of the different activities usually performed; at a later time, the module has to recognize occurring activities based on previously created models. All the information about resident habits and preferences is provided as the output of this module. This information is then used by the "Energy and Comfort Management Module" to monitor and act on appliances and devices according to a precise scheduling algorithm based on user profiles and on activities performed. Indeed, each activity can be associated with a specific set of appliances that are turned on or adjusted accordingly. Table 1 includes the details of the most typical home appliances, along with their probability to be available at home in Italy [4]. Based on the activity recognition and/or prediction results, the appliances can be scheduled to improve energy savings and/or users' comfort [4]. The decisions taken by this last module can operate on the system automatically, via commands sent to actuators, or they can be translated in useful advice sent to the user through some interface. All the described modules are integrated into an intelligent device that oversees the data storage and the control of the building, either locally or in the cloud [39,40].

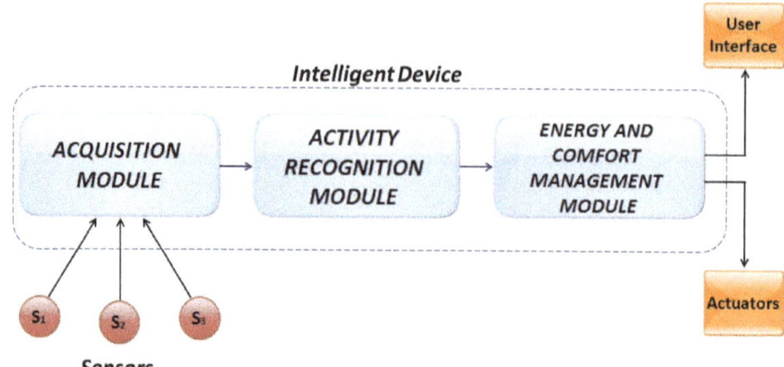

Figure 1. Integration of the activity recognition system inside a BECM system.

Table 1. Characteristic parameters of the most typical home appliances [4].

Name	Power [Wh]	Mean Execution Time [min]	Probability to Have It
Fridge/freezer	70	Always on	100%
Lighting	40	Always on when someone is at home	100%
PC/laptop	50	150	95%
TV	30	210	100%
Game console	90	120	5%
Hair dryer	1500	15	100%
Iron	1100	20	100%
Microwave oven	1000	90	52%
Washing machine	600	130	86%
Dishwasher	400	160	34%
Clothes dryer	1300	90	8%
Electric oven	2000	15	53%
HVAC	1000	Always on when someone is at home	31%
Water heater	2000	Always on	50%

As an explanatory example, suppose that according to their profile, a user usually wakes up, then has a coffee watching TV, and later takes a shower with the bathroom heater on. Furthermore, suppose that the *wake up* activity is detected when the *turn on the bedroom light* event is identified (i.e., the light sensor inside the user bedroom detects some light) after the *sleep* activity occurred. Since the BECM system knows that the following activities are *have a coffee*, *watch TV* and *take a shower*, it can increase the user's comfort by: turning on the coffee machine as soon as the user wakes up, turning on the TV right after the coffee is made, and at the same time turning on the water and bathroom heater so that the water and room are warm when the user goes to the bathroom. Alternatively, the BECM system can choose to switch water and bathroom heater on earlier if the energy cost is lower, based on predictions about the usual morning routine of the user.

Figure 2 expresses the relationship between sensors, events, and activities. An event corresponds to a change in the state of a sensor. Each sensor has several possible states, depending on the significant values it measures; this is the case, for example, of a contact on a door, which has two different states (e.g., *OPEN*, *CLOSE*): an event is registered whenever the contact changes its state from *OPEN* to *CLOSE* and vice versa. On the other hand, a smart meter monitoring a washing machine can have multiple states: according to the measured power consumption, its related state can correspond to *OFF* or to any wash cycle. Accordingly, an event is registered whenever the smart meter changes its state.

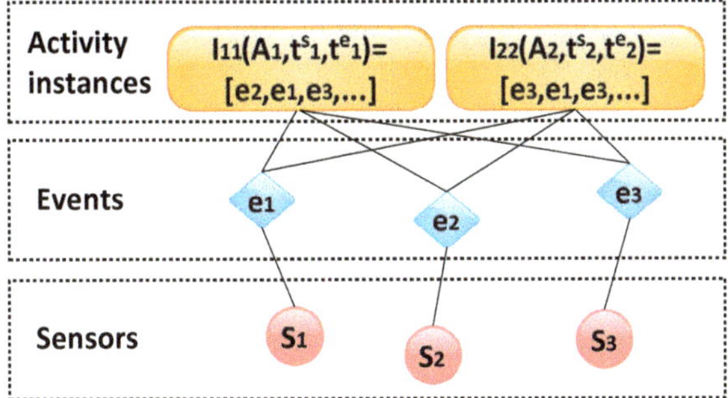

Figure 2. Example of activities and relevant subdivision into actions and observations, in connection with related sensors.

Let $\mathcal{E} = \{e_i\}$ be the set of events that can be detected by sensors inside a house. An event is then defined as the transition between possible states of the corresponding sensor monitoring that particular situation. According to this vision, activities are composed of several events. Even though the events that characterize an activity remain basically the same, their order may change when considering two different observations of the same activity. For example, when preparing a meal one can open the fridge to take the ingredients and then put a pan on the stove, or the same activity can be done in the opposite order. For this reason, a generic activity $\mathcal{A}_j$ is modeled as the set of events that are usually observed when the reference user performs it. Accordingly, an instance $\mathcal{I}_{jk}(\mathcal{A}_j, t^s_k, t^e_k)$ of activity $\mathcal{A}_j$ is defined as the array of events that are observed from the time t^s_k the activity started to the time t^e_k the activity finished. In the following sections, the process used to recognize the activities based on the observed events will be described in detail.

4. Activity Recognition Algorithm

The set of events $\mathcal{E}$ is the basic set of information needed to understand what users are doing, and as seen previously they are strongly connected to sensors activation. The proposed system is then based on recognizing activities performed in smart environments from sequences of collected sensor readings. The choice of which sensor types to use leads to different models and algorithms for solving activity recognition problems because of different kinds of data, produced according to the type of sensor that has to be manipulated in order to extract important information.

As seen in Section 2, the solutions for monitoring people in their home include: 1. cameras, 2. wearable sensors and 3. different kinds of ambient sensors. The first solution gives the type of information with the highest accuracy, but it requires the heaviest process to manipulate images and videos. Moreover, some people find problematic the idea of having cameras in their homes. Wearable sensors give information about the physical state of a person, so it is possible to understand simple actions, but it is harder to recognize more complex and complicated activities. Besides, there could be problems if the users forgot to wear them. Ambient sensors produce data with a low level of semantic information, but choosing only non-intrusive binary sensors is a better option for experiments in real life so that there is no need for people to remember to always wear wearable sensors or to be monitored with cameras.

The flowchart in Figure 3 expresses and clarifies all the steps of the proposed activity recognition algorithm that are described in detail in the following lines. The algorithm is constituted by two main phases: the training phase, during which the characteristics and statistics that describe how the observed user performs the activities are created; the running phase, where the events are observed and processed so that activities can be recognized by the classification module.

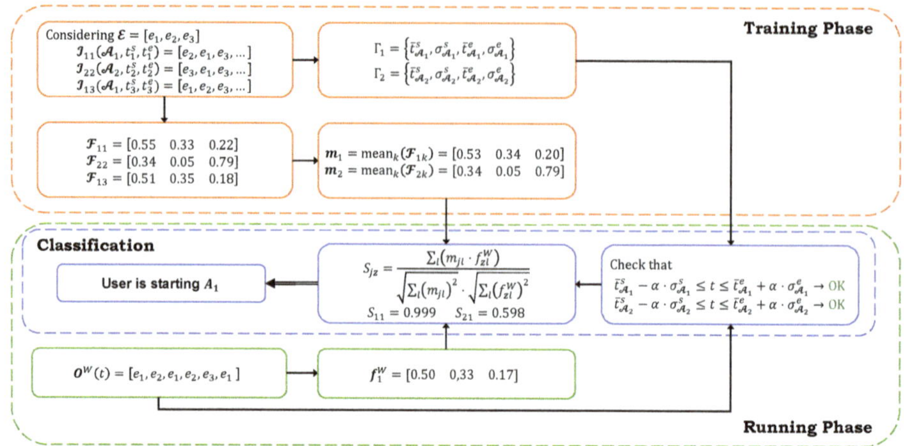

Figure 3. Flowchart of the steps for the activity recognition algorithm.

4.1. Training Phase

During the training phase, each activity instance is observed within a time window $\mathcal{O}^A$ included in $\{t_k^s, t_k^e\}$ and is defined by the sequence of sensor events, i.e., sensors that have changed their state within the considered window. For each observed activity, a feature vector $\mathcal{F}_{jk}(\mathcal{I}_{jk}) = [f_{1jk}, f_{2jk}, \ldots, f_{ijk}, \ldots]$ is computed with the rates of event occurrences for each sensor that is the number of events related to one specific sensor with respect to the total number of events observed considering all the sensors within the time window $\mathcal{O}^A$. Then, for each type of activity $\mathcal{A}_j$, a model vector $\boldsymbol{m}_j = \text{mean}_k(\mathcal{F}_{jk}) = [\bar{f}_{1jk}, \bar{f}_{2jk}, \ldots, \bar{f}_{ijk}]$ is defined such that the rates of event occurrences of its sensors is the average rate for all the observed instances associated with the same activity. This model vector $\boldsymbol{m}_j$ is representative about the probability that a sensor is connected to every activity. When an event associated with a sensor is counted for a certain number of times during the observed sequence, it is possible to understand which activity is statistically more probable.

From raw data acquired by sensors, feature vectors that can be analyzed to perform classification of activities are constructed. Every activity is strongly connected to a specific group of sensors that change their states during a definite time span, as described in the previous section.

Furthermore, statistics Γ_j about relevant conditions that can be associated with the performed activity $\mathcal{A}_j$ are evaluated and stored by the system. Indeed, activities are often performed within the same time window (e.g., sleeping, preparing meals). Therefore, Γ_j includes the following statistics about $\mathcal{A}_j$: the average starting and ending times $\bar{t}^s_{\mathcal{A}_j}$ and $\bar{t}^e_{\mathcal{A}_j}$, and their standard deviations $\sigma^s_{\mathcal{A}_j}$ and $\sigma^s_{\mathcal{A}_j}$. If there is more than one time window, statistics are generated for each observed time window.

4.2. Running Phase

After the probabilistic model is obtained, the system must recognize the activities performed by evaluating which the most likely to be happening is. To this aim, a sensor-based windowing implementation is considered [9]. This approach consists of dividing a sequence of incoming events into subsequences using an observation window $\mathcal{O}^W(t)$ starting at time t, which contains a certain number of events equal to the size $\mathcal{W}$ of the aforesaid window. Another possible approach would have been the time-based windowing, which consists of segmenting the incoming sequence of events using a window of fixed temporal length [30]. This second approach is mostly used when analyzing data coming from sensors such as gyroscope or accelerometer, because there is a constant amount of data over time, while with binary sensors there could be moments without any sensor readings and some stall phases. Since the proposed system uses this type of binary sensors, a sensor-based window has been chosen. The result is that every sensor is treated as a feature and is associated with a particular activity based on its distribution probability to be in a sequence that is labeled with the name of that activity. This is done by implementing a Naïve Bayes Classifier (NBC) [41]. This type of classifier is based over the Bayes' theorem of independence between features, so that the model is constructed to find, for each activity $\mathcal{A}_j$, the probability p that given a certain number of features, i.e., the set of events, the class being observed is $\mathcal{A}_j$:

$$p(\mathcal{A}_j|\mathcal{E}) \tag{1}$$

Accordingly, for each sequence of events observed in the observation window $\mathcal{O}^W$, a feature vector $\boldsymbol{\mathcal{F}}_z^W$ is computed similarly to the previous vector $\boldsymbol{\mathcal{F}}_{jk}$ of the training phase.

Finally, in the classification phase, the sequences of observed events are classified based on their probability to belong to a given activity. To this, the possible activities to be associated with the observed sequence of events are first filtered based on statistics Γ_j: only the activities that are usually observed within a time window that includes the current observation window's starting time t are considered in the following step. Accordingly, the following condition needs to be fulfilled:

$$\bar{t}^s_{\mathcal{A}_j} - \alpha \cdot \sigma^s_{\mathcal{A}_j} \leq t \leq \bar{t}^s_{\mathcal{A}_j} + \alpha \cdot \sigma^s_{\mathcal{A}_j} \tag{2}$$

where α is a weighting factor that adjusts the time window to be considered. Finally, the cosine similarity S_{jz} between model vector $\boldsymbol{m}_j$ and feature vector $\boldsymbol{f}_z^W$ is calculated with the equation below over the remaining activities, to evaluate which the more likely to be observed is:

$$S_{jz} = \frac{\sum_i (m_{ji} \cdot f_{zi}^W)}{\sqrt{\sum_i (m_{ji})^2} \cdot \sqrt{\sum_i (f_{zi}^W)^2}} \tag{3}$$

The observed activity is then labeled as the activity that corresponds to the highest cosine similarity.

5. Reference Use Case

The algorithm for modeling the activities and then discovering what the resident is doing is implemented and tested using the Aruba real-word dataset from the CASAS smart environment project of the Washington State University [37]. The data were collected from one smart apartment provided with motion sensors, contact sensors in the doors or cabinets and temperature sensors. Figure 4 shows the house plan of the apartment and the exact position of every sensor in the rooms. The description provided by the CASAS project did not give information about the specifics of the used sensor. Table 2 explains the number of sensors per type placed in each room. The values provided by motion and contact sensors are Boolean, whereas the ones provided by temperature sensors are numbers. There are two more sensors not listed in this table, one motion sensor and one contact sensor because they are not located in a specific room, but they are linked to the entrance of the house.

To correctly evaluate the correlation between the sets of events and the observed user's activities, without interference from other people, a dataset with only one resident living in the home was considered. The events decoded by these sensors are significant for recording elementary actions that people are performing, for example, door sensors are easily associated with opening and closing medical cabinet, food storage, or the entrance door, while with motion sensors it is possible to monitor the presence of the resident in one room and the proximity with a specific object or piece of furniture. The aggregation of these elementary actions defines one activity of interest.

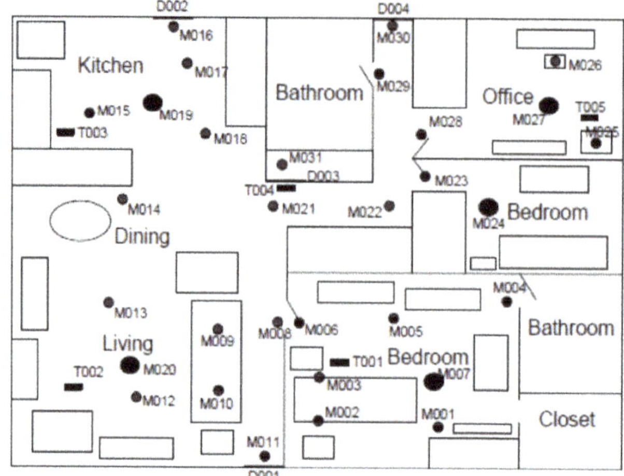

Figure 4. House plant of the apartment for the Aruba dataset [37].

Table 2. Number of sensors per type in every room of the apartment.

	Motion Sensors	Contact Sensors	Temperature Sensors
Kitchen	5	1	1
Bathroom 1	2	1	-
Office	4	-	1
Dining	1	-	1
Bedroom 1	2	-	-
Living	7	1	1
Bedroom 2	7	-	1
Bathroom 2	1	-	-
Closet	-	-	-

The gathered data are presented with information about the date and time of every sensor event registered, the id of the activated sensor with its value and the beginning or end of each activity that is monitored. The dataset has the structure presented in Table 3. In the dataset, 10 different activities performed by the resident are noted. Table 4 shows the details of the number of times each activity appears in the dataset, as indicated by the user. The "Relax" activity is the one that the user has denoted as the activity occurring while staying in the living room, and it involves the set of sensors arranged in that room, as shown in the house map (Figure 4). The "Work" activity is the one performed in the office room and involving the specific group of sensors placed in that area. Lastly, the "Housekeeping" activity involves a great number of all the sensors of the house, due to the intrinsic dynamism of this type of activity. Sensors detect even the activities that are not registered, that correspond to "Other activity" with no label in the dataset. Since they cannot be classified accurately, this has been ignored in the proposed framework.

Table 3. Data extracted from the Aruba dataset [37].

Date	Time	Sensor ID	Sensor Value	Activity
4 November 2010	09:56:22.785482	M018	ON	
4 November 2010	09:56:23.801652	M017	ON	
4 November 2010	09:56:26.467399	M019	ON	
4 November 2010	09:56:27.334395	M018	OFF	Meal Preparation end
4 November 2010	09:56:34.362031	M018	ON	
4 November 2010	09:56:37.729204	M020	ON	
4 November 2010	09:56:38.776094	M018	OFF	
4 November 2010	09:56:40.172391	M020	OFF	
4 November 2010	09:56:41.831135	M014	ON	Eating begin
4 November 2010	09:56:56.043362	M014	OFF	
4 November 2010	09:57:15.209217	M014	ON	
4 November 2010	09:57:16.412611	M014	OFF	

Table 4. Activities and Statistics of the Aruba dataset.

Activity	Number of Occurrences
Meal Preparation (MP)	1606
Relax (Rel)	2910
Eating (Eat)	257
Work	171
Sleeping (Sleep)	401
Wash Dishes (WD)	65
Bed to Toilet (BTT)	157
Enter Home (EH)	431
Leave Home (LH)	431
Housekeeping (HK)	33

Part of the dataset has been first used to train the system, whereas another part has been used to test it, by simulating the running phase. Most parts of the monitored activities have a long representation in terms of sequences of events. Only three of them generally are concluded in less than 15 sensor events, and they are the activities of "Enter Home", "Leave Home" and "Bed to Toilet Transition". The others are more variable and can last longer. Because of this variability, it is difficult to find a common size $\mathcal{W}$ to consider for the observation window $\mathcal{O}^{\mathcal{W}}$. The size of this window is essential because for the activity recognition problem the algorithm evaluates the sensors' events that occur within this window. Based on that, the most probable activity being performed is found by searching for the minimum distance between the modeled feature vector and the new instances that are happening at the moment and that has to be classified, as defined in the previous Section.

6. Performance Evaluation

To evaluate the algorithm, an assessment of the classification accuracy that shows the percentage of correctly classified sequences of events for each class is used. The accuracy is obtained by observing the number of times an activity is correctly labeled, compared to all the occurrences of that activity in the dataset. Accordingly, the accuracy for activity l is expressed by

$$Accuracy_l = \frac{T_l}{T_l + F_l} \tag{4}$$

where T_l and F_l are respectively the number of times the activity was correctly and erroneously labeled.

The test is performed considering a training time of 2 months, one week of test data from the dataset and taking a window of $\mathcal{W} = 10$ sensor events as the sequence that must be classified. The choice of the window is done by taking into account that there are three activities among those indicated in Table 4 that are shorter than the others. These activities often last for several events included between 5 and 15. Furthermore, it is preferable that activities are recognized as early as possible so that relevant management actions can be promptly started. Nevertheless, choosing a window that is too small could lead to errors when longer activities, which are the most numerous, are being performed, because the algorithm frequently confuses possible activities that are similar.

6.1. Accuracy of the Activity Recognition Algorithm

To first evaluate the accuracy of the recognition algorithm by itself, the system was first run excluding the filtering with respect to the Γ_j statistics, as it is proposed in [9]. With the current choices, simulation results achieve an average accuracy of 70.43%. The results are presented in Figure 5a where the confusion matrix with the accuracy percentages of the classification for all the activities is presented.

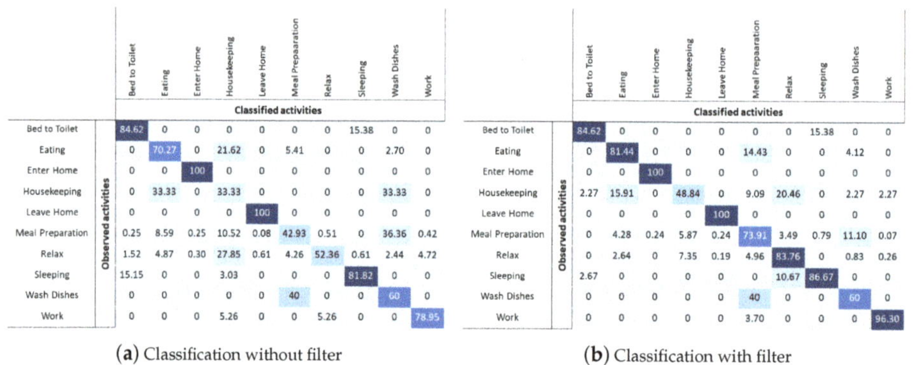

(a) Classification without filter

Observed activities \ Classified activities	Bed to Toilet	Eating	Enter Home	Housekeeping	Leave Home	Meal Prepaaration	Relax	Sleeping	Wash Dishes	Work
Bed to Toilet	84.62	0	0	0	0	0	0	15.38	0	0
Eating	0	70.27	0	21.62	0	5.41	0	0	2.70	0
Enter Home	0	0	100	0	0	0	0	0	0	0
Housekeeping	0	33.33	0	33.33	0	0	0	0	33.33	0
Leave Home	0	0	0	0	100	0	0	0	0	0
Meal Preparation	0.25	8.59	0.25	10.52	0.08	42.93	0.51	0	36.36	0.42
Relax	1.52	4.87	0.30	27.85	0.61	4.26	52.36	0.61	2.44	4.72
Sleeping	15.15	0	0	3.03	0	0	0	81.82	0	0
Wash Dishes	0	0	0	0	0	40	0	0	60	0
Work	0	0	0	5.26	0	0	5.26	0	0	78.95

(b) Classification with filter

Observed activities \ Classified activities	Bed to Toilet	Eating	Enter Home	Housekeeping	Leave Home	Meal Prepaaration	Relax	Sleeping	Wash Dishes	Work
Bed to Toilet	84.62	0	0	0	0	0	0	15.38	0	0
Eating	0	81.44	0	0	0	14.43	0	0	4.12	0
Enter Home	0	0	100	0	0	0	0	0	0	0
Housekeeping	2.27	15.91	0	48.84	0	9.09	20.46	0	2.27	2.27
Leave Home	0	0	0	0	100	0	0	0	0	0
Meal Preparation	0	4.28	0.24	5.87	0.24	73.91	3.49	0.79	11.10	0.07
Relax	0	2.64	0	7.35	0.19	4.96	83.76	0	0.83	0.26
Sleeping	2.67	0	0	0	0	0	10.67	86.67	0	0
Wash Dishes	0	0	0	0	0	40	0	0	60	0
Work	0	0	0	0	0	3.70	0	0	0	96.30

Figure 5. Confusion matrices of the activity recognition algorithm proposed in [9] (**a**), which corresponds to a classification without the proposed filter, and of the proposed algorithm that includes the statistics-based filter (**b**).

It is evident that almost half of the times there are errors discerning the "Meal Preparation" activity from the "Wash Dishes" activity. This is due to the fact that both activities are performed in the kitchen and involve the same sensors. The reason for the frequent errors between the "Relax" and "Housekeeping" activities is explainable remembering how this second activity was described in the previous Section 5. Due to the fact that in several cases the sensors involved are the ones placed in the living room, the algorithm often confuses the two activities.

Results improve consistently when adding the filter before the cosine similarity calculation, as it is shown in Figure 5b. Indeed, in this second case, the algorithm gave an average accuracy of 81.14% and it is possible to observe how all the activities are better recognized. This is even more evident in Figure 6, where the comparison of the percentage accuracy for every class between the two examined cases is presented.

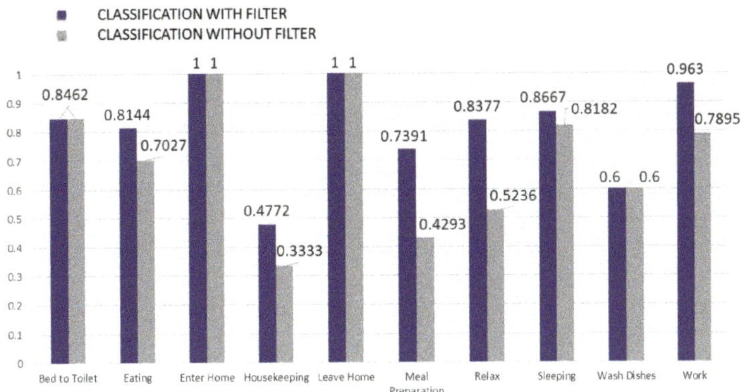

Figure 6. Comparison between activity recognition accuracy in percentage with and without the filter. The classification without the filter corresponds to the algorithm in [9].

6.2. *Accuracy Results for Different Sizes of the Observation Window* $\mathcal{O}^W$

Table 5 shows the accuracy results for the different sizes W of the observation window with the indication of the 95% confidence interval for the three examined cases. Changing the size of the observation window to 15 and 20 consecutive sensor events, the performance of the algorithm are quite better, giving results of average accuracy equal to 87.02% and 84.09% respectively, making the choice of the size window of 15 sensor events the preferable one. This is due to the reason explicated before, for which longer activities are more easily observed with bigger windows while shorter activities are usually less frequent and affect the performance less significantly. Results also confirm that performance shrinks when considering observation windows that are longer than the shortest activities, i.e., those that have less than 15 events as mentioned at the beginning of this Section. The confidence interval is narrower for smaller windows than for bigger windows, because in these cases there is a greater number of samples, due to the fact that the same instance is divided into more parts compared to the use of a bigger window.

Table 5. Accuracy and confidential interval using different size of observation window W.

	Accuracy	95% Confidence Interval
***W size* = 10**	81.14%	±1.43%
***W size* = 15**	87.02%	±1.56%
***W size* = 20**	84.09%	±1.77%

6.3. *Accuracy Results for Different Training Periods*

In Figure 7 the different overall accuracy values obtained with increasing training days, from 2 weeks to 3 months, are presented. The horizontal bars represent the 95% confidence interval with values that go from ±1.95% for 2 weeks of training to ±1.68% for 3 months of training.

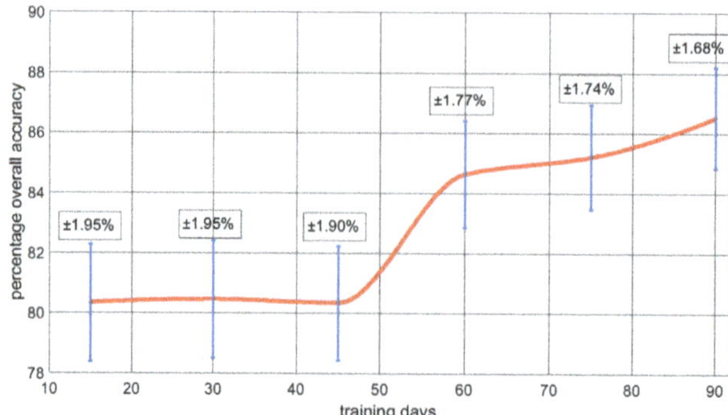

Figure 7. Percentage overall accuracy for different training days.

The increase in the performance from 2 months to 3 months is not as prominent as the increasing noticed during fewer days of training. Analyzing this result, the conclusion is that it is probably better to have a training phase that does not last too long so that is not necessary to wait a long time period to get results. Instead, it is more important to evaluate the need for a new training phase after a while, in order to check if there have been changes in the habits of the user after some time. Furthermore, the various activities under consideration are not carried out with the same frequency every day or every week. This explains why, with a training set of a few days, the accuracy obtained is so stable and low, due to the fact that during that time some activities have only a few samples that could be used for making the models, making it harder to recognize those same activities later in the running phase.

6.4. Prediction Results for Subsequent Activities

The current activity can be used to predict the activities that are going to be performed in the next future. For this purpose, an evaluation of the probability of transition from an activity in a column to an activity in a row is reported in Table 6, whereas Table 7 shows the probabilities that the activity "C" is happening given the fact that activities "A" and "B" occurred. In the last table, only some of the results are presented, i.e., the ones with the greatest probability value for each combination of the first two activities. When two consecutive activities indicated in the tables turn out to be the same repeated activity, it is due to the fact that between these two instances of the same activities there was a gap in which the user performed some actions that were unknown and that were part of the "Other activity" group that was not considered during these tests as it was specified in Section 5. The last column reports the average time duration expressed in minutes of the sequence of the 3 activities indicated as activity "A", "B" and "C". This information could be useful for future works for making prediction and consideration of possible sequences of activities over a long time.

Table 6. Transition Probability of Activities.

	BTT	Eat	EH	HK	LH	MP	Rel	Sleep	WD	Work
Bed to Toilet	0.017	0	0	0	0	0	0	0.41	0	0
Eating	0	0.14	0	0.030	0.12	0.02	0.034	0.007	0.61	0.068
Enter Home	0	0.015	0	0.061	0.26	0.095	0.041	0.035	0	0.136
Housekeeping	0	0	0	0	0.042	0.0036	0.006	0	0	0.025
Leave Home	0	0	1	0	0	0	0	0	0	0
Meal Preparation	0	0.75	0	0.121	0.009	0.374	0.237	0	0	0.11
Relax	0	0.09	0	0.58	0.403	0.344	0.644	0.523	0.356	0.348
Sleeping	0.98	0	0	0	0.0035	0.14	0.003	0.042	0	0
Wash Dishes	0	0.005	0	0.091	0.017	0.003	0.0197	0	0.017	0.042
Work	0	0.01	0	0.121	0.059	0.003	0.015	0.014	0.017	0.27

Table 7. Conditional probability for a sequence of three activities.

Activity A	Activity B	Activity C	Probability (C ǀ A ∩ B)	Duration
Bed to Toilet	Sleeping	Meal Preparation	0.769	249.4
Sleeping	Bed to Toilet	Sleeping	0.983	241.86
Sleeping	Sleeping	Meal Preparation	0.75	246.6
Eating	Meal Preparation	Eating	0.455	27.4
Eating	Wash Dishes	Relax	0.667	24.97
Enter Home	Eating	Relax	0.667	50.87
Enter Home	Meal Preparation	Meal Preparation	0.448	74.31
Meal Preparation	Eating	Relax	0.36	33.86
Meal Preparation	Relax	Relax	0.624	74.12
Meal Preparation	Work	Meal Preparation	0.462	74.31
Relax	Wash Dishes	Relax	0.667	74.124
Relax	Meal Preparation	Relax	0.462	74.31
Relax	Sleeping	Bed to Toilet	0.597	275.26
Relax	Work	Relax	0.30	253.6
Wash Dishes	Relax	Relax	0.781	84.5
Wash Dishes	Meal Preparation	Eating	0.667	24.79
Wash dishes	Work	Relax	0.60	59.45
Housekeeping	Meal Preparation	Relax	0.50	61.06
Housekeeping	Relax	Relax	0.462	87.12
Leave Home	Enter Home	Meal Preparation	0.365	247.55
Work	Housekeeping	Leave Home	0.624	39.7
Work	Sleeping	Bed to Toilet	0.75	260.7

7. Conclusions and Future Works

This paper focuses on the problem of users' activity recognition inside Smart Buildings to support BECM systems, by exploiting the acquisitions made by sensors deployed inside the building. To this aim, a system that analyses the user behavior to profile it and later perform early recognition and prediction of users' activities is presented. The proposed system is a sensor-based activity recognition system that analyzing raw data from binary sensors arranged in an apartment can model users' activities and understand known behaviors. In addition to the data collected by sensors, other significant information, provided by statistics Γ_j related to relevant external conditions that have been observed to be correlated (e.g., time of the day), are incorporated in the system, allowing the obtaining of better results than the simple analysis case of sensors data. The system is proved to enable activity recognition with an accuracy of more than 80% after only 10 sensor events are registered. Furthermore, the prediction of the following activity is achieved with an accuracy of about 60%.

It should be mentioned, however, that there are some limitations to the proposed method. As already highlighted explaining the obtained results, there are some problems related to the fixed size of the observation window, which hold the performance of the algorithm depending on the activity examined. A wider window is better to recognize long activities but, on the other hand, a longer time is required to start the assignment of an activity. Another important aspect concerns the period of the training phase. A longer time span gives better accuracy results, with more available samples to better model each activity, but it leads to exaggerated waiting times to first get any results. Lastly, this approach was thought and tested only for consecutive activities performed by a single user. Some changes must be evaluated for an extension to more than one resident and for recognize concurrent and interleaved activities.

Future works will be focused on improving the system accuracy by including statistics of other relevant external conditions that can be correlated, such as the weather, or the fact that it is a working day or not. Furthermore, other devices, such as smartphones, will be included to enlarge the number of sensors available for the analysis. The introduction of measurements coming from personal devices is expected to provide a more thorough insight into users' habits. Moreover, the system will be expanded to consider cases with more residents and to recognize different contemporary actions. Finally, the proposed system will be first tested using commercial software, to be later included on a real BECM system scenario, so that the actual convenience on energy cost savings and the quality of experience perceived by users can be assessed.

Author Contributions: Conceptualization, F.M., V.P.; validation: F.M.; writing–draft preparation, F.M., V.P., D.G.; writing–review and editing, F.M., V.P., D.G.

Funding: This work was partially supported by MIUR, within the Smart Cities framework (Project CagliariPort2020, ID: SCN 00281), by the Italian Ministry of Economic Development (MiSE, Project INSIEME, HORIZON 2020, PON 2014/2020 POS. 395), by "Fondazione di Sardegna" within the research project "SUM2GRIDS—Solutions by mUltidisciplinary approach for intelligent Monitoring and Management of power distribution GRIDS"—Convenzione triennale tra la Fondazione di Sardegna e gli Atenei Sardi, Regione Sardegna—L.R. 7/2007 annuity 2017—DGR 28/21 of 17 May 2015, within the project Design and Implementation of a Novel Hybrid Energy Storage System for Microgrids, which is funded by the Sardinian Regional Government (Regional Law no. 7, 7 August 2007) under the Grant Agreement no. 68 (Annuity 2015), and within the project LEAPH – anaLytics and data Enrichment plAtform for Pharma and pHarmacy owner, funded by the Sardinian Regional Government under the Sardinian POR FESR 2014-2020 (ID: RICERCA_1C-124).

Conflicts of Interest: The authors declare no conflicts of interest.

References

1. Minoli, D.; Sohraby, K.; Occhiogrosso, B. IoT considerations, requirements, and architectures for smart buildings—Energy optimization and next-generation building management systems. *IEEE Int. Things J.* **2017**, *4*, 269–283. [CrossRef]
2. Dounis, A.I.; Caraiscos, C. Advanced control systems engineering for energy and comfort management in a building environment—A review. *Renew. Sustain. Energy Rev.* **2009**, *13*, 1246–1261. [CrossRef]
3. Shafie-Khah, M.; Siano, P. A stochastic home energy management system considering satisfaction cost and response fatigue. *IEEE Trans. Ind. Inf.* **2017**, *14*, 629–638. [CrossRef]
4. Pilloni, V.; Floris, A.; Meloni, A.; Atzori, L. Smart home energy management including renewable sources: A qoe-driven approach. *IEEE Trans. Smart Grid* **2016**, *9*, 2006–2018. [CrossRef]
5. Yang, L.; Yan, H.; Lam, J.C. Thermal comfort and building energy consumption implications—A review. *Appl. Energy* **2014**, *115*, 164–173. [CrossRef]

6. Shoreh, M.H.; Siano, P.; Shafie-Khah, M.; Loia, V.; Catalão, J.P. A survey of industrial applications of Demand Response. *Electr. Power Syst. Res.* **2016**, *141*, 31–49. [CrossRef]
7. Pooranian, Z.; Abawajy, J.; Conti, M. Scheduling distributed energy resource operation and daily power consumption for a smart building to optimize economic and environmental parameters. *Energies* **2018**, *11*, 1348. [CrossRef]
8. Rasheed, M.; Javaid, N.; Ahmad, A.; Khan, Z.; Qasim, U.; Alrajeh, N. An efficient power scheduling scheme for residential load management in smart homes. *Appl. Sci.* **2015**, *5*, 1134–1163. [CrossRef]
9. Krishnan, N.C.; Cook, D.J. Activity recognition on streaming sensor data. *Pervasive Mob. Comput.* **2014**, *10*, 138–154. [CrossRef] [PubMed]
10. Shaikh, P.H.; Nor, N.B.M.; Nallagownden, P.; Elamvazuthi, I.; Ibrahim, T. A review on optimized control systems for building energy and comfort management of smart sustainable buildings. *Renew. Sustain. Energy Rev.* **2014**, *34*, 409–429. [CrossRef]
11. Marinakis, V.; Doukas, H. An advanced IoT-based system for intelligent energy management in buildings. *Sensors* **2018**, *18*, 610. [CrossRef] [PubMed]
12. Marinakis, V.; Doukas, H.; Karakosta, C.; Psarras, J. An integrated system for buildings' energy-efficient automation: Application in the tertiary sector. *Appl. Energy* **2013**, *101*, 6–14. [CrossRef]
13. Wang, L.; Wang, Z.; Yang, R. Intelligent multiagent control system for energy and comfort management in smart and sustainable buildings. *IEEE Trans. Smart Grid* **2012**, *3*, 605–617. [CrossRef]
14. Du, P.; Lu, N. Appliance commitment for household load scheduling. *IEEE Trans. Smart Grid* **2011**, *2*, 411–419. [CrossRef]
15. Collotta, M.; Pau, G. Bluetooth for Internet of Things: A fuzzy approach to improve power management in smart homes. *Comput. Electr. Eng.* **2015**, *44*, 137–152. [CrossRef]
16. Koseleva, N.; Ropaite, G. Big data in building energy efficiency: Understanding of big data and main challenges. *Procedia Eng.* **2017**, *172*, 544–549. [CrossRef]
17. Stimmel, C.L. *Big Data Analytics Strategies for the Smart Grid*; Auerbach Publications: New York, NY, USA, 2016.
18. Al Nuaimi, E.; Al Neyadi, H.; Mohamed, N.; Al-Jaroodi, J. Applications of big data to smart cities. *J. Int. Serv. Appl.* **2015**, *6*, 25. [CrossRef]
19. Attal, F.; Mohammed, S.; Dedabrishvili, M.; Chamroukhi, F.; Oukhellou, L.; Amirat, Y. Physical human activity recognition using wearable sensors. *Sensors* **2015**, *15*, 31314–31338. [CrossRef]
20. Nitti, M.; Stelea, G.A.; Popescu, V.; Fadda, M. When Social Networks Meet D2D Communications: A Survey. *Sensors* **2019**, *19*, 396. [CrossRef]
21. Lu, Y.; Wei, Y.; Liu, L.; Zhong, J.; Sun, L.; Liu, Y. Towards unsupervised physical activity recognition using smartphone accelerometers. *Multimed. Tools Appl.* **2017**, *76*, 10701–10719. [CrossRef]
22. Liu, Y.; Nie, L.; Liu, L.; Rosenblum, D.S. From action to activity: Sensor-based activity recognition. *Neurocomputing* **2016**, *181*, 108–115. [CrossRef]
23. Nitti, M.; Popescu, V.; Fadda, M. Using an IoT platform for trustworthy D2D communications in a real indoor environment. *IEEE Trans. Netw. Serv. Manag.* **2019**, *16*, 234–245. [CrossRef]
24. Diraco, G.; Leone, A.; Siciliano, P. People occupancy detection and profiling with 3D depth sensors for building energy management. *Energy Build.* **2015**, *92*, 246–266. [CrossRef]
25. Klein, L.; Kwak, J.Y.; Kavulya, G.; Jazizadeh, F.; Becerik-Gerber, B.; Varakantham, P.; Tambe, M. Coordinating occupant behavior for building energy and comfort management using multi-agent systems. *Autom. Constr.* **2012**, *22*, 525–536. [CrossRef]
26. De Kok, R.; Mauri, A.; Bozzon, A. Automatic Processing of User-Generated Content for the Description of Energy-Consuming Activities at Individual and Group Level. *Energies* **2019**, *12*, 15. [CrossRef]
27. Viard, K.; Fanti, M.P.; Faraut, G.; Lesage, J.J. An event-based approach for discovering activities of daily living by hidden Markov models. In Proceedings of the 2016 15th International Conference on Ubiquitous Computing and Communications and 2016 International Symposium on Cyberspace and Security (IUCC-CSS), Granada, Spain, 14–16 December 2016; pp. 85–92.
28. Rohini, P.; RajKumar, R. A New Approach to Behavioral Reasoning in Smart Homes using DVSM Algorithm. *Int. J. Adv. Res. Comput. Sci. Softw. Eng.* **2014**, *4*, 334–341.
29. Rashidi, P.; Cook, D.J.; Holder, L.B.; Schmitter-Edgecombe, M. Discovering activities to recognize and track in a smart environment. *IEEE Trans. Knowl. Data Eng.* **2011**, *23*, 527–539. [CrossRef] [PubMed]

30. Thomas, B.L.; Cook, D.J. Activity-aware energy-efficient automation of smart buildings. *Energies* **2016**, *9*, 624. [CrossRef]
31. Roy, N.; Misra, A.; Cook, D. Ambient and smartphone sensor assisted ADL recognition in multi-inhabitant smart environments. *J. Ambient Intell. Humaniz. Comput.* **2016**, *7*, 1–19. [CrossRef]
32. Chiang, Y.T.; Hsu, K.C.; Lu, C.H.; Fu, L.C.; Hsu, J.Y.J. Interaction models for multiple-resident activity recognition in a smart home. In Proceedings of the 2010 IEEE/RSJ International Conference on Intelligent Robots and Systems, Taipei, Taiwan, 18–22 October 2010; pp. 3753–3758.
33. Wilson, D.H.; Atkeson, C. Simultaneous tracking and activity recognition (STAR) using many anonymous, binary sensors. In *International Conference on Pervasive Computing*; Springer: Berlin/Heidelberg, Germany, 2005; pp. 62–79.
34. Yala, N.; Fergani, B.; Fleury, A. Feature extraction for human activity recognition on streaming data. In Proceedings of the 2015 International Symposium on Innovations in Intelligent SysTems and Applications (INISTA), Madrid, Spain, 2–4 September 2015; pp. 1–6.
35. Shoaib, M.; Bosch, S.; Incel, O.D.; Scholten, H.; Havinga, P.J. Complex human activity recognition using smartphone and wrist-worn motion sensors. *Sensors* **2016**, *16*, 426. [CrossRef]
36. Nef, T.; Urwyler, P.; Büchler, M.; Tarnanas, I.; Stucki, R.; Cazzoli, D.; Müri, R.; Mosimann, U. Evaluation of three state-of-the-art classifiers for recognition of activities of daily living from smart home ambient data. *Sensors* **2015**, *15*, 11725–11740. [CrossRef] [PubMed]
37. Cook, D.J. Learning setting-generalized activity models for smart spaces. *IEEE Intell. Syst.* **2010**, *27*, 32–38. [CrossRef] [PubMed]
38. Han, J.; Choi, C.S.; Lee, I. More efficient home energy management system based on ZigBee communication and infrared remote controls. *IEEE Trans. Consum. Electron.* **2011**, *57*, 85–89.
39. Botta, A.; De Donato, W.; Persico, V.; Pescapé, A. Integration of cloud computing and internet of things: A survey. *Future Gener. Comput. Syst.* **2016**, *56*, 684–700. [CrossRef]
40. Josep, A.D.; Katz, R.; Konwinski, A.; Gunho, L.; Patterson, D.; Rabkin, A. A view of cloud computing. *Commun. ACM* **2010**, *53*, 50–58.
41. Kaur, G.; Oberai, E.N. A review article on Naive Bayes classifier with various smoothing techniques. *Int. J. Comput. Sci. Mob. Comput.* **2014**, *3*, 864–868.

Article

Greedy Algorithm for Minimizing the Cost of Routing Power on a Digital Microgrid

Zhengqi Jiang [1], Vinit Sahasrabudhe [1], Ahmed Mohamed [2], Haim Grebel [1] and Roberto Rojas-Cessa [1,*]

1 Department of Electrical and Computer Engineering, New Jersey Institute of Technology, Newark, NJ 07102, USA
2 Department of Electrical Engineering, City College of City University of New York, New York, NY 10031, USA
* Correspondence: rojas@njit.edu

Received: 22 June 2019; Accepted: 30 July 2019; Published: 9 August 2019

Abstract: In this paper, we propose the greedy smallest-cost-rate path first (GRASP) algorithm to route power from sources to loads in a digital microgrid (DMG). Routing of power from distributed energy resources (DERs) to loads of a DMG comprises matching loads to DERs and the selection of the smallest-cost-rate path from a load to its supplying DERs. In such a microgrid, one DER may supply power to one or many loads, and one or many DERs may supply the power requested by a load. Because the optimal method is NP-hard, GRASP addresses this high complexity by using heuristics to match sources and loads and to select the smallest-cost-rate paths in the DMG. We compare the cost achieved by GRASP and an optimal method based on integer linear programming on different IEEE test feeders and other test networks. The comparison shows the trade-offs between lowering complexity and achieving optimal-cost paths. The results show that the cost incurred by GRASP approaches that of the optimal solution by small margins. In the adopted networks, GRASP trades its lower complexity for up to 18% higher costs than those achieved by the optimal solution.

Keywords: digital microgrid; power grid; integer linear programming; routing energy; distributed energy resources; Dijkstra algorithm; integer linear programming

1. Introduction

The power grid infrastructure has been dramatically evolving from the traditional electrical grid to the smart grid to make it more manageable and efficient [1–4]. Moreover, the interest in integrating renewable resources into the grid to decrease greenhouse gas emissions into the atmosphere continues to increase. It is no surprise that the next-generation smart grid will use optimized electricity generation and transmission infrastructure, provide elastic electricity distribution, reduce peaks in power usage, and sense and react to power outages by incorporating the Internet and information technologies along with intelligent control algorithms [5,6].

Microgrids are becoming a functional and building block of the next-generation power grids [1]. They enable local control for the distribution of power and configurations tailored to satisfy local needs. These features are critical for improving the reliability and resilience of power distribution in the occurrence of failures or natural disasters [7]. The flexibility of microgrids to embrace one or multiple energy resources, configure one or multiple paths for redundancy, and the incorporation of communications to monitor and control the configuration of distribution legs enable such features [8]. Moreover, microgrids have access to the main power line in addition to the local energy resources. Therefore, microgrids can operate in grid-connected and islanded modes [9]. In a grid-connected mode, DERs may increase supply capacity, and in islanded mode, DERs may energize critical loads as needed during extenuating circumstances, such as in emergencies.

Several works propose the concept of the digital grid (DG) as means to fulfill the requirements of the next-generation power grid [10–19]. There are several approaches to the DG. They all share the use of internet protocol (IP) addresses to identify lines and grid equipment, such as power routers, to make them be able to communicate and participate in a the management and the distribution of power. The DG not only enables a robust operation and a seamless integration of renewable energy sources into a grid or microgrid, but it also leverages the incorporation of energy storage. The use of energy storage habilitates the aggregation of power generated by different sources and the supply of specific amounts of power to specific loads. The transmission of data in the Internet have inspired the development of the DG [20–23].

The DG supplies energy in a specific amount and delivers it to a unique destination, such as a piece of grid equipment, link interconnecting a substation, energy storage unit, or the final consuming load. In the DG, the energy carries both the IP address of the recipient and information of the amount of power delivered [14,24,25]. The transmission of the recipient address and the amount of granted power may be performed by either power lines that also transmit data or by a parallel data network. On a higher operating layer, a request-grant protocol is performed between loads and sources to schedule the delivery of power [14,24,25]. In such a protocol, loads issue a request for the needed amount of power, and one or several sources grant such requests and supply the requested amount of power. The DG concept is particularly suited for managing a microgrid, which becomes a digital microgrid (DMG). A service provider, acting as a DMG controller, may exercise centralized management for a DMG with numerous sources.

Digitization of power (i.e., the supply of finite and exact amounts of power to a specific destination or interconnection leg) makes it possible to route power from one or multiple energy sources to a load of a DMG. This routing of power, however, is difficult to realize on a traditional microgrid. In a DMG, the distribution network, formed by the interconnected transmission lines and cables, may yield more than one path for a source to supply power to a load. Shinabo [26] considers a similar feature, although the focus is the optimization of the power generation by multiple generators. Here, each generator is independent and interconnected by a power router [27]. However, this approach suffers from both limited scalability and uncontrollable power delivery because the power router only sets the connection or disconnection of lines rather than controlling the amount of power delivery.

However, the efficiency of a DMG may be affected by the costs of power distribution. In this paper, we are interested in minimizing the total cost, which we can model as power loss. We define routing here as the selection of the smallest-cost-rate links and as an optimization problem that calls for integer linear programming (ILP) as the solving approach [28]. The cost-rate is the rate that each unit of power costs to distribute (i.e., cost= cost rate * power).

Because microgrids may count with multiple distributed energy resources (DERs), the routing problem includes a matching between a load and one or multiple DERs to satisfy the load's power demand. We show the optimal selected paths on different power networks, including different IEEE test buses, where each of them includes a different number of energy sources and loads. While this optimal method works well in the networks tested in this paper, we envision a power network with a vast number of loads for which the high complexity of ILP would make the selection of paths infeasible. Because the optimal ILP method is NP-hard, we propose the greedy smallest-cost-rate path first (GRASP), a heuristic approach but with lower complexity than ILP. GRASP matches the loads to sources and finds the smallest-cost-rate paths to route power in an iterative and greedy approach, and it uses the Dijkstra algorithm as the path searching mechanism [29]. We analyze the cost of GRASP and compare it with that of the optimal method. The discrepancy between the two approaches is the cost associated with the reduction of algorithm complexity. We show that GRASP approaches the optimal solution by 18% additional cost, but with a reduced computation complexity.

We organize this paper as follows. Section 2 introduces an optimal routing and source assignment method and GRASP. Section 3 presents the evaluation results of the optimal method and GRASP for different distribution networks, including IEEE test buses. This section also includes a practical example of the application of GRASP. Section 4 presents our conclusions.

2. Methods: Optimal and GRASP

In a conventional microgrid, the delivery of power takes place over a single shared bus, which directly connects the power source to each load. However, a large number of DERs in a DMG ensemble a distributed power generation. Examples of such DERs are rooftop solar panels, wind turbines, and conventional fossil-fuels-based generators. We model the set of interconnections formed by generators, grid nodes (or power routers), and loads as a graph where edges represent distribution links and nodes represent sources, power routers, or loads. Power routers may be multi-leg devices; they enable building a distribution network with multiple distribution paths. Therefore, a DMG may distribute power to loads through disjoint lines. Here, several loads may share a bus and the supply from many DERs but they may use diverse paths to interconnect to DERs. The DMG uses a power router to transfer energy from one link (or smaller grid) to another [30]. Such a power router has multiple inputs and outputs, or legs. Such routers can interconnect multiple segments and form an extended microgrid. The end topology of the microgrid would be a meshed network [11]. Figure 1 shows the architecture of a power router. This power router comprises a power path, which delivers energy to loads, and a data path, which transmits the information to monitor the demand and supply of energy. The inputs and outputs of the power router handle data and power. Here, energy is aggregated or partitioned by the router to supply amounts that match those requested.

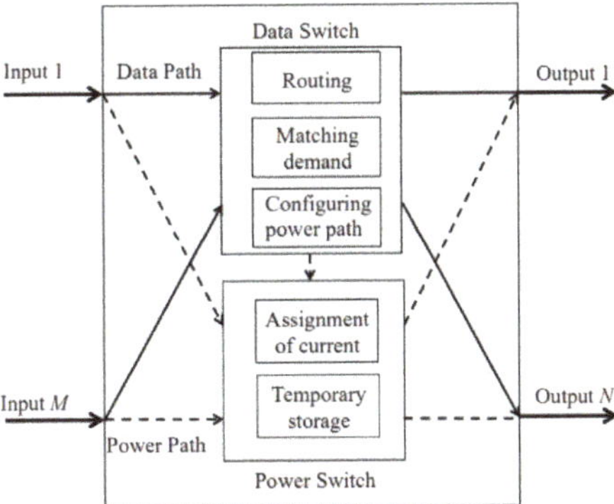

Figure 1. Architecture of a power router. Reproduced from [14]. Publisher: IEEE.

In this paper, we focus on operating expenses (OPEX) of the use of power routers in a DMG. We consider losses on the transmission of power to be the associated costs. We consider two sources of losses: transmission lines and power routers. The lines may have a constant loss or one proportional to the energy transmitted. We adopt the latter one. The routers have power losses inherent to the transfer of energy from one input to an output, associated with the power router operation. Also, because the microgrid may be extended, there may be a cost associated with these extensions, as worst-case scenarios. For simplicity, and without losing generality, we associate these cost-rates to each link. Because some links and paths may incur costs as power flows through them, it is of interest to minimize the cost (or overhead) of power distribution in the DMG. We also address the practical issue of how links can be interconnected to enable multiple supply paths between sources and loads, and are also useful for fault-tolerant purposes. A router with many legs may enabled us to set up a complex network that includes redundant paths in the microgrid.

Moreover, the presence of multiple DERs allows the supply to a load by a single or multiple energy sources and an energy source to supply single or multiple loads. Thus, matching loads to sources was also a process included in the routing of energy in a DMG. The minimum cost routes include the assignment of sources to loads through minimum cost paths.

2.1. Problem Definition

The following example depicts the path and source assignment problem in the supply of power requests in a DMG. Figure 2a shows an example of a simple DMG. This network has two sources (nodes 1 and 2), two loads (nodes 3 and 4), and two power routers (nodes 5 and 6). The two power routers provide a path with links that share the power transmitted to both loads. Each source has a supply capacity indicated by the labels (0.5R units for node 1 and R units for node 2), and the number on each link indicates the cost per unit of transmitted power and the capacity (i.e., the amount of power that the link can carry) as (p, q), respectively. Each load requests an amount of power, as indicated by the labels beside the node.

In this example, node 3 requests R units of power, and node 4 requests 0.5R units. Figure 2b shows a path for transmitting power from node 2 (source) to node 3. This path has a low cost but leaves a load request unsatisfied as there is no other path to connect to node 1 and because the shared link (link from node 5 to node 6) has reached its capacity with the flow from node 2 to node 3. The figure shows the flow that supplies power to Node 3 as a solid blue line. Figures 2c,d shows the flow that supplies node 4 as a dashed red line. Figure 2c shows that the selected paths satisfy all requests, but the total cost is high (7R). Figure 2d shows a solution where the two loads are also satisfied, and the cost is the lowest. This solution also shows an example of a source supplying two loads (node 2) and a load receiving energy from two sources (node 3).

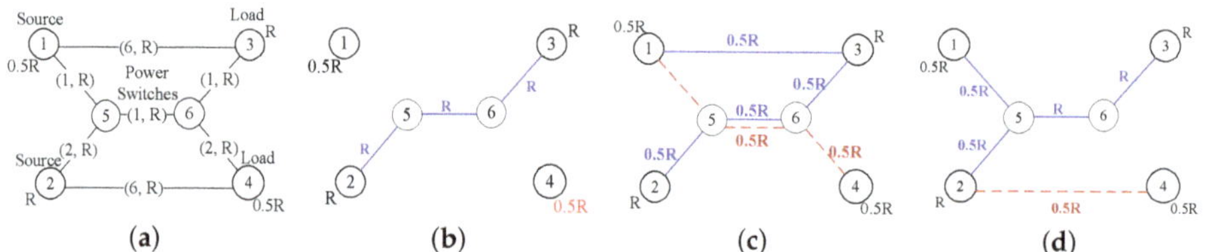

Figure 2. Example of problem definition of source and path selection on a small digital microgrid (DMG) model: (a) network model with paths, two sources and two loads; (b) Low cost path but unsatisfied requests. Total cost is 4R; (c) satisfied requests but high routing cost. Total cost is 7R; and (d) Lowest cost path and satisfied requests. Total cost is 6.5R.

As the example shows, the optimal routing of power in a DMG comprised of (a) the selection of the lowest-cost path to supply each load, (b) the selection of sources that supply energy according to their generation capacities, and (c) the consideration of the link transmission capacities.

2.2. Optimal Solution: Integer Linear Programming Formulation

We model the selection of paths with the smallest cost as an optimization problem to minimize the energy cost on the distribution of energy. The cost and losses in the distribution network of the DMG were linear and expressed with integer numbers, therefore, ILP is suitable for solving this optimization problem. We first modeled the DMG as a distribution network. Consider $G = (V, E)$ be a graph with two special vertices: a source s and a sink t. Here, u, where $u \in V$, is a node in the set of nodes V and edge (i, j), such that $(i, j) \in E$, in the set of edges E. Every edge has a capacity $cap(i, j) \geq 0$ and a cost or weight associated with that edge. The capacity and cost are two main features of the edges in the graph that indicate the maximum capacity of the link and cost per power unit (kW) transmitted, respectively. One or multiple flows may use one edge. A flow is a real-valued function on edges that occupies a portion of the power capacity of a link, and it is indicated as $f(i, j)$. In practice, one can set the transmitted capacity as the amount of power (or current) that the link can carry and that the source can provide. The properties of a flow are as follows:

- Capacity constraints: $f(i,j) \leq cap(i,j)$. If $f(i,j) = cap(i,j)$, we say that edge (i,j) is saturated.
- Flow conservation (at each node except the sources and sinks): $\sum f(i,j) = 0, \forall i \in V - \{s,t\}$. This property indicates that all flows that enter a transport (router) node also exit that node. Here, we assume that the switching of power has insignificant energy overhead.

The set of ILP equations describe the network model, the selection of paths and the matching between loads and sources. This set of equations consists of an objective function and a set of constraints. The objective function is defined as follows:

$$min \sum_{(i,j)\in E} d_{ij}x_{ij}. \tag{1}$$

Subject to:

$$0 \leq s_i, \forall i \in V_s \tag{2}$$

$$0 \leq t_j, \forall j \in V_t \tag{3}$$

$$\sum_{j:(i,j)\in E} x_{ij} - \sum_{j:(j,i)\in E} x_{ji} = s_i, \forall i \in V_s \tag{4}$$

$$\sum_{i:(i,j)\in E} x_{ij} - \sum_{i:(j,i)\in E} x_{ji} = t_j, \forall j \in V_t \tag{5}$$

$$\sum_{j:(i,j)\in E} x_{ij} - \sum_{j:(j,i)\in E} x_{ji} = 0, \forall i \in V - V_s - V_t \tag{6}$$

$$0 \leq x_{ij} \leq cap(i,j), \forall (i,j) \in E, \tag{7}$$

where x_{ij} is the amount of power the generator transmits to the load, d_{ij} is the cost per unit of power edge (i,j) carries, s_i and t_j are the amount of power source i generates and sink j requests, respectively. V_s is a subset of V that contains all the source nodes. Similarly, V_t contains all the sink nodes. Here, (1) defines the cost minimization function, (2) and (3) are constraints corresponding to the amount of power generated by the sources and required by the sinks, (4) and (5) describe the conservation of energy flow of the power sources and sinks, respectively, (6) describes the conservation of energy flow of the transport nodes, and (7) shows the capacity constraint for each link. The complexity of the optimal method for the selection of the smallest cost path is $O(n^{(2|V|+2)}(|V|a)^{(|V|+1)(|V|+2)})$ where n is the largest number of legs of a network node, $a = max\{|(i,j)|, |V_s|, |V_t|\}$; the largest link cost, source capacity, and load demand, and $|V|$ is the number of nodes in the network.

2.3. Greedy Smallest-Cost-Rate Path First (GRASP) Algorithm

Although the optimal method converges to an optimal solution for energy allocation in the presence of multiple energy sources and loads, the method is NP-hard, so that its complexity is prohibitively high for a network with a considerable number of nodes [31]. GRASP aims to match sources and loads and, in combination, find the smallest cost-rate paths in between, restricted to the source and link capacities, and the losses associated to links. By using the cost rates, GRASP aims to obtain a small routing cost. It should be noted that this smallest cost rate may not be the smallest overall cost but the use of this parameter reduces the complexity of finding the paths in comparison with the optimal method. The algorithm operates in an iterative and greedy approach, and its objective is also described by the objective function and constraints in Section 2.2. GRASP finds the k shortest cost-rate paths, one per each load-source pair and matches loads (demands) to as many sources as needed with an aggregate capacity generation that is able to satisfy the power demands of the loads.

There are other constraints in this problem: (a) the amount of power that the selected path can carry was set by either (1) the link with the smallest of power capacity; (2) the capacity of the power source and; (3) the amount of power the customer requests. (b) The selection of a path cannot select a congested link (i.e., a link carrying an amount of power equal to its capacity) for accommodating another request. Because different sources may partially fulfil the power requested by a load, the number of paths needed to supply a demanding load may be many. A routing algorithm must select the sources needed to satisfy a load demand and whose path to the source(s) incurs in the minimum transmission costs. This property is a significant difference between routing in a DMG and that in computer networks.

GRASP works as follows:

Step 1. Find all smallest cost-rates paths between for all load and sources.

Step 2. Select the path with the smallest cost rates among those in step 1, while considering the amount of power (generation) capacity of the source, the link capacities, and the amounts of power requested by the loads.

Step 3. Reserve the portion of the path and source capacity for the routed flow and consider that portion of the request satisfied (if not all).

Step 4. Decrease the capacity of each link on the selected path and source by the minimum of the requested amount and that granted. If the portion of available power left on one of the links of the selected path or sources reaches zero, delete this link or source from the set obtained in step 1.

Step 5. Increase the amount of power granted to the load, and if the request is full satisfy it, consider this load satisfied.

Step 6. Repeat the process, starting from step 1, until all the customers receive all the requested power or not paths or source remains available for supplying the requested amount of power.

The pseudocode of this algorithm, presented in Algorithm 1, shows a greater level of detail.

Therefore, it is easy to see that the complexity of GRASP is the complexity of the smallest-cost-rate link search or $O(|E|log|V|)$ [32] times the number of loads times the number of sources, or $O(N_s\ N_L\ |E|log|V|)$, where N_s is the number DERs, N_L is the number of loads whose power requests are satisfied, and $|E|$ is the number of links in the network.

Algorithm 1 Greedy smallest-cost-rate path first (GRASP).

```
procedure CALCULATE TOTAL COST
    Initialize a connected graph G(V, E)
    total_cost ← 0
    path ← smallest-cost-rate paths for source-load pairs
    cost ← costs of path(i) in path
    loop:
    if not path and one or more loads request power then
        return No paths; End
    if no load request power then
        return total_cost
    else
        min_cost ← min(cost) % here cost is the cost rate
        p ← the i-th path with cost(i) == min_cost
        cap_s ← power capacity of source on path p
        req_l ← power demands of load on path p
        link_p ← capacity of all the links on path p
        min_link_p ← minimum path capacity p
        min_tran ← min( cap_s, req_l, min_link _p)
        total_cost ← total_cost + min_tran * cost(i)
        cap_s ← cap_s - min_tran
        req_l ← req_l - min_tran
        link_p(i) ← link_p(i) - min_tran ∀ links of path p
        update path andcost :
        s ← source of path p
        l ← load of path p
        link_min ← link with min capacity of path p
        if min_tran == cap_s then
            delete all the paths including source s
            delete the costs from cost
        if min_tran == req_l then
            delete all the paths connected to load l
            delete the costs of cost
        if min_tran == min_link_p then
            if link_min in graph G then
                delete link_min
            find p_new with the smallest-cost rate cost_new
            add p_new and cost_new to path and cost
        goto loop
```

3. Results and Analysis

Here, we evaluate the total cost incurred by the optimal method and GRASP. We show the application of both methods on two different networks. One was a simple network and the other was National Science Foundation (NSFNET). While NSFNET was not a power network, was is an example of a network with a different number of paths between two nodes (as in computer networks). It also shows how a power distribution bus may look after the adoption of power routers. We also consider different IEEE test buses, which found their use in various power tests. Some of these networks had a small number of power sources and loads, and others had a large number of both, thus presenting different scenarios. These networks offered different examples of the application of multi-leg nodes and a variety of paths.

3.1. Example DMG Networks

3.1.1. A Simple Network

First, we consider a simple network, as Figures 3 and 4 show. These figures show the cost and capacity link assignment and source capacity for case 1. In the remainder of this paper, we show diagrams for case 1 of the considered networks. In this network, there are three power sources, nodes 1–3, depicted in green color. Each source node had a generation capacity of 35 kW, 45 kW, and 20 kW, respectively, resulting in a total generation of 100 kW available to the two loads; nodes 10 and 11, depicted in yellow color. Each load requested 50 kW. We indicate both the cost (e.g., power loss) per unit of power transmitted through it and *power capacity* of each link on the network.

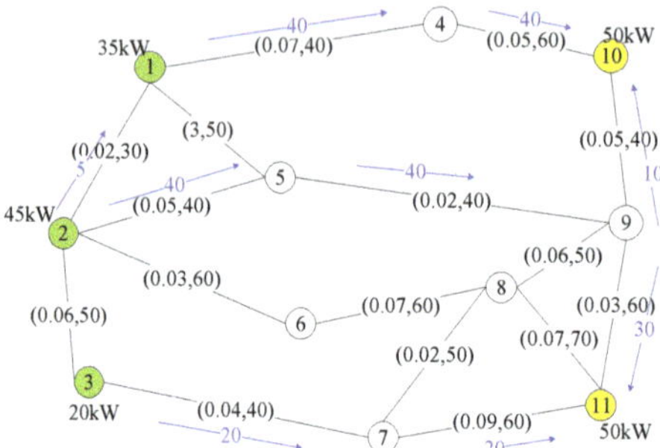

Figure 3. Paths selected by the optimal method on a simple network.

We consider three test cases per network. We assigned an arbitrary cost and capacity per link in case 1. We modified the costs and link capacity of the links in case 2 to test the selection of a different path. We changed the amount of power capacity of sources and demand (requests) for driving the loads to match with different sources, leading to a varied selection of paths, in case 3.

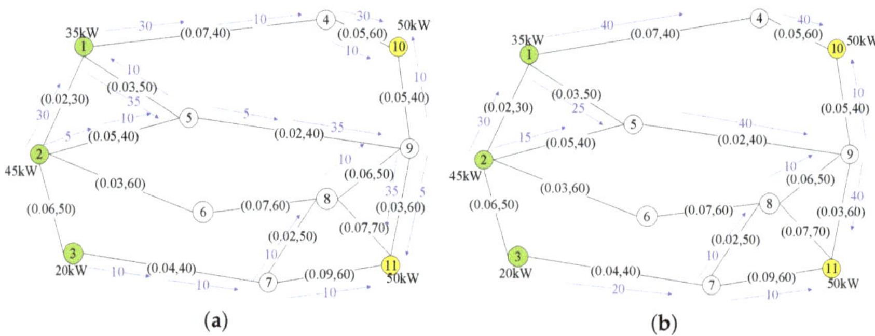

Figure 4. Paths selected by greedy smallest-cost-rate path first (GRASP) on a simple network. (**a**) Paths selected by GRASP on a simple network indicating each power flow. (**b**) Paths selected by GRASP on a simple network indicating the aggregated power.

In general, the GRASP algorithm greedily selects the smallest-cost-rate path of all the source-load pairs in each iteration. As shown in line 4 of the pseudocode, the smallest-cost-rate paths of all source-load pairs are pre-calculated and cached in a list path. For each iteration of the loop, in line 13, path p, which is the smallest one amount the paths in path, will be selected and the source of this path is assigned to the load of this path. To show how the proposed GRASP algorithm works, we took the simple network as an example. As shown in Figure 4a, the blue arrows show the selected paths by using GRASP. First, GRASP calculates the smallest-cost-rate path of all the source-load pairs. It selects the path $1 \rightarrow 5 \rightarrow 9 \rightarrow 11$, where $a \rightarrow b$ means a link selected from node a to node b, that has a cost rate equal to 0.08. Then source 1 transmits 35 kW of power through as limited by its capacity. Then, it selects $2 \rightarrow 5 \rightarrow 9 \rightarrow 11$ as the second smallest-cost-rate path, to transmit 5 kW because this is the maximum power that could be transmitted form node 5 to 9 as 35 kW were already reserved on this link. GRASP keeps going in this way until all the power demands are satisfied.

A path from a source to a load can be represented by the path cost, which is the sum of link costs, and the path capacity, which is the capacity of the link with the smallest power capacity of all links comprising the path. The total cost is the sum of incurred flow costs on each link:

$$C_t = \sum_{(i,j) \in E} d_{ij} x_{ij}, \tag{8}$$

where C_t is the total cost, d_{ij} is the cost per unit of power transmitted on edge (i,j), and x_{ij} is the amount of power transmitted on edge (i,j).

Figures 3 and 4 also show the selected paths and the amounts of power granted over each selected path, as blue arrows, selected by the optimal method and GRASP, respectively. As Figure 4a shows, some power flows share some of the links in the selected paths as they are associated with small costs. Figure 4b shows the total power flowing on each link. The power on the links is the sum of the individual power per flow shown in Figure 4. We use this representation of results in the remainder of this paper. In the network, the total cost of routing power from all the sources to all the loads was 12.7 kW. However, the total cost achieved by GRASP was 13.5 kW or an increase of 6.3% over the optimal solution. We also performed a test for cases 2 and 3 (we do not show the figures for these two cases because of the limited space). In case 3, the power capacity of nodes 1–3 were 40 kW, 40 kW, and 20 kW, respectively, and the amount of power requested by nodes 10 and 11 was 70 kW and 30 kW, respectively. Figure 5 shows the comparisons of the total cost incurred by the optimal method and GRASP for each the three cases. As the figure shows, GRASP achieved a cost of 8.33% above the optimal costs as the worst case of the tested scenarios.

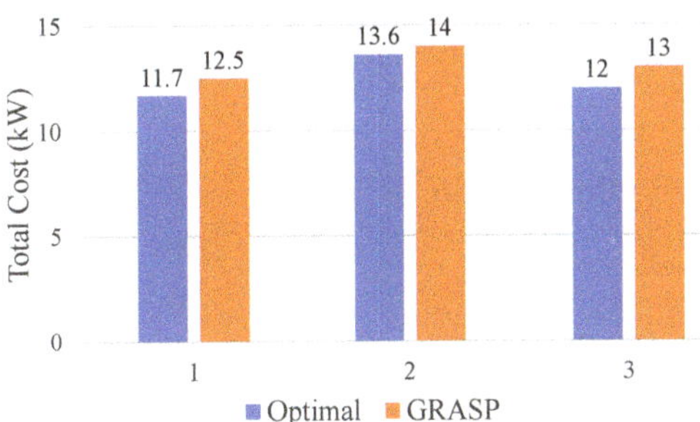

Figure 5. Cost comparison between the optimal solution and GRASP on the simple network.

We also performed route calculations on more complex networks, such as NSFNET [33], IEEE14 test [34] and IEEE39 test buses [35,36]. In this section, we present the total cost recurred by GRASP and the optimal method. Similarly, as in the simple network discussed above, we consider three different cases for each network.

3.1.2. NSFNET

This network is the result of a program of coordinated and evolving projects sponsored by the National Science Foundation (NSF) that was initiated in 1985 to support and promote advanced networking among U.S. research and education institutions. Figures 6 and 7 show this network. This network had three sources and two loads, as in the simple network, but it also had a larger number of links and a higher node degree. The increased number of links in this network enabled a wider selection of independent paths (i.e., paths with fewer or no shared link) of source-load pairs. Particularly to this network, the cost of each link is directly proportional to the geographical distance between two nodes (each node in NSFNET is a U.S. city) and the capacity of each link was set arbitrarily.

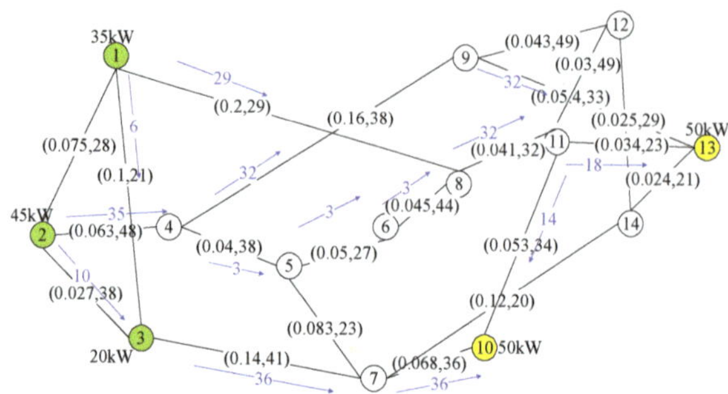

Figure 6. Optimal path selection and source-load matching on the NSFNET network.

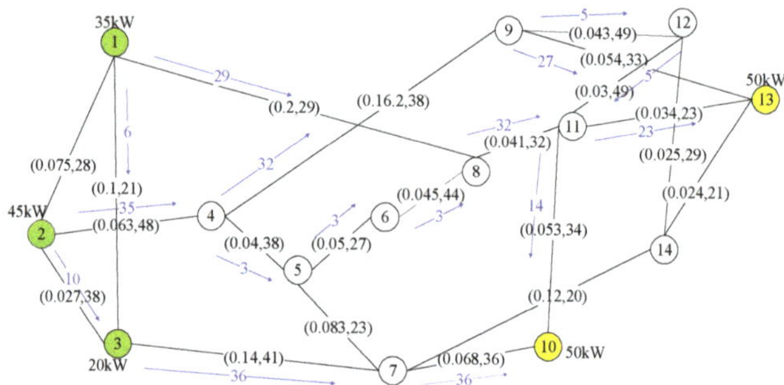

Figure 7. Path selection and source-load matching by GRASP on the NSFNET network.

Figures 6 and 7 show the results of case 1 in NSFNET as resolved by the optimal method and GRASP, respectively. Figure 8 shows the overall cost results obtained by these two methods. These results indicate that GRASP fell 3.9% short of achieving the optimal solution by an increase of about 12.6 kW in case 2. Similarly to the simple network, case 2 of the NSFNET network used different link costs and link power capacities from case 1, while case 3 used different power source capacities and demands of power. Case 2 then is used to observe different route selection from case 1 as link costs are modified, and case 3 may lead to the use of different paths but triggered by a different distribution on power generation and demand. From these two cases, case 2 draws the larger discrepancy between these two methods. However, the differences between the optimal solution and GRASP in the NSFNET network are smaller than those achieved in the simple network. A possible cause of this improvement is the larger number of routing options provided by the larger number of links and node order in NSFNET.

Considering the complexity of NSFNET, the matching of sources and loads and the selected distribution paths in this network is also represented by a power routing table, as Tables 1 and 2. These tables correspond to the path selection presented in Figures 6 and 7, as obtained by the optimal method and GRASP, respectively. These tables show the amounts of power assigned to the different links. In each table, x_{ij} means the amount of power transmitted from node i to node j. A positive x_{ij} means that the power flows from node i to node j and a negative x_{ij} means that the power flows from node j to node i (opposite direction). Also, if $x_{ij} = 0$, there is no power flow between nodes i and j.

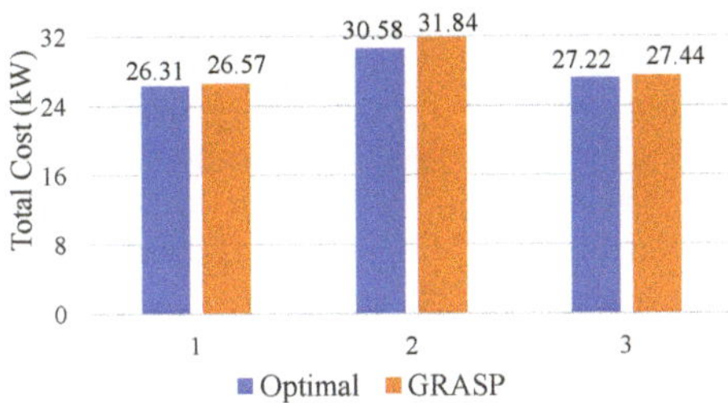

Figure 8. Cost comparison between the optimal method and GRASP on the NSFNET network.

Table 1. Power assigned to the links of the NSFNET network by using the optimal method.

Energy Routing Table (kW)													
Node	1	2	3	4	5	6	7	8	9	10	11	12	13
1	0	0	6	0	0	0	0	29	0	0	0	0	0
2	0	0	10	35	0	0	0	0	0	0	0	0	0
3	-6	−10	0	0	0	0	36	0	0	0	0	0	0
4	0	−35	0	0	3	0	0	0	32	0	0	0	0
5	0	0	0	−3	0	3	0	0	0	0	0	0	0
6	0	0	0	0	−3	0	0	3	0	0	0	0	0
7	0	0	−36	0	0	0	0	0	0	36	0	0	0
8	−29	0	0	0	0	−3	0	0	0	0	32	0	0
9	0	0	0	−32	0	0	0	0	0	0	0	0	32
10	0	0	0	0	0	0	−36	0	0	0	−14	0	0
11	0	0	0	0	0	0	0	−32	0	14	0	0	18
12	0	0	0	0	0	0	0	0	0	0	0	0	0
13	0	0	0	0	0	0	0	0	−32	0	−18	0	0

Table 2. Power assigned to the links of the NSFNET network by GRASP.

Energy Routing Table (kW)													
Node	1	2	3	4	5	6	7	8	9	10	11	12	13
1	0	0	6	0	0	0	0	29	0	0	0	0	0
2	0	0	10	35	0	0	0	0	0	0	0	0	0
3	−6	−10	0	0	0	0	36	0	0	0	0	0	0
4	0	−35	0	0	3	0	0	0	32	0	0	0	0
5	0	0	0	−3	0	3	0	0	0	0	0	0	0
6	0	0	0	0	−3	0	0	3	0	0	0	0	0
7	0	0	−36	0	0	0	0	0	0	36	0	0	0
8	−29	0	0	0	0	−3	0	0	0	0	32	0	0
9	0	0	0	−32	0	0	0	0	0	0	0	5	27
10	0	0	0	0	0	0	−36	0	0	0	−14	0	0
11	0	0	0	0	0	0	0	−32	0	14	0	−5	23
12	0	0	0	0	0	0	0	0	−5	0	5	0	0
13	0	0	0	0	0	0	0	0	−27	0	−23	0	0

3.1.3. IEEE14 bus

The IEEE14 bus system [34] represents a simple approximation of the American Electric Power system as of February 1962. This test bus has 14 buses, five generators (i.e., sources), and 11 loads.

However, to stress the performance of GRASP in this network, we assume only three loads from those in the bus. In this way, we increase the search space to match the loads to the sources as the number of sources is much larger than the number of loads. Figures 9 and 10 show the matched pairs and resulting paths by the two considered schemes. The data in the figures indicate the amount of power provided by the sources and requested by the loads and the cost and capacity of each link of case 1.

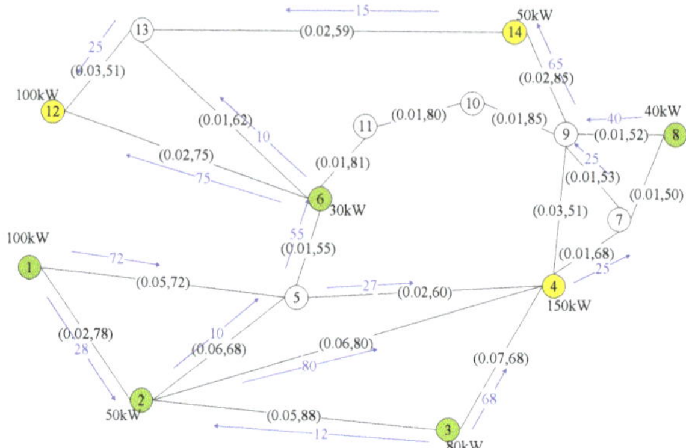

Figure 9. Optimal path selection and source-load matching on the IEEE14 bus.

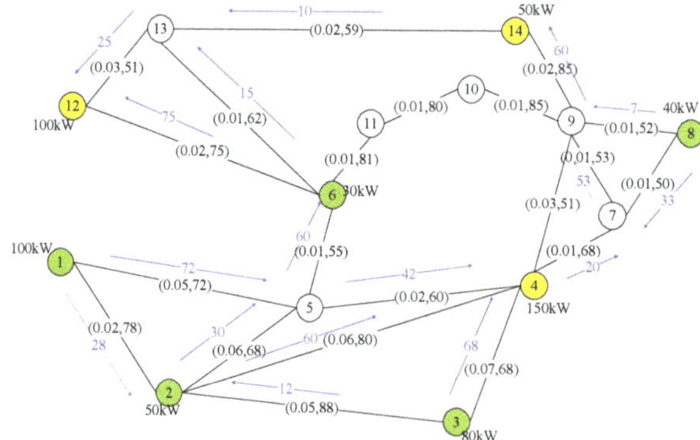

Figure 10. Path selection and source-load matching performed by GRASP on the IEEE14 bus.

The results in this figures show that the total cost of the paths selected by GRASP is comparable to that achieved by the optimal method, as Figure 11 shows. The largest discrepancy in this scenario was 18.0%, which was the additional cost incurred by the selections GRASP make.

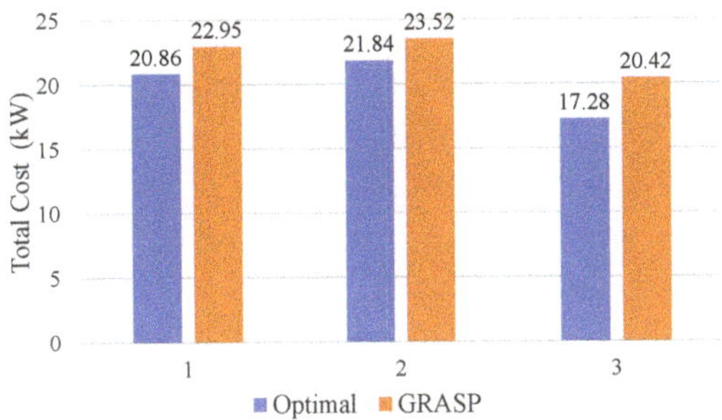

Figure 11. Comparison of power costs achieved by the optimal method and GRASP on the IEEE14 bus system.

3.1.4. IEEE39 bus

This was a large bus, it had 39 buses, 46 transmission lines, 10 power sources, and 19 loads. This network had two nodes, nodes 31 and 39, represented in blue color, that performed a combined functional mode; each node generated and spent energy. These two nodes played both roles; a power source and a load, at the same time.

Figures 12 and 13 show the results obtained by the optimal method and GRASP on this test bus. With the larger number of sources and customers, the IEEE39 test bus showed a much more complex scenario than the three previous network topologies. Here, loads had a larger number of choices of sources to select, but the number of links and node order is equal to or smaller than those considered in NSFNET. It was expected that there were fewer choices for routing power in this test bus than in NSFNET and at the same time, more flows needed to be routed by the larger number of loads in this test bus.

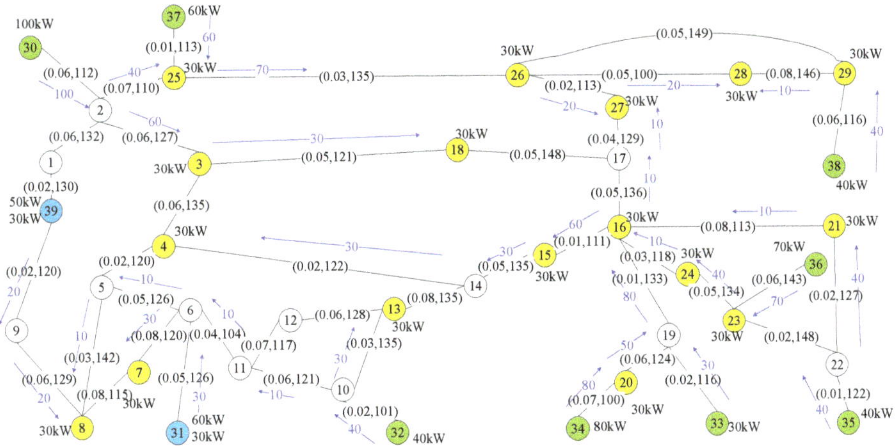

Figure 12. Optimal path selection and source-load matching on the IEEE39 bus.

Figure 14 shows a comparison of the different cases tested on this network. This graph shows that GRASP achieves 9% as the most substantial additional cost in comparison to the optimal solution in case 3. Figure 15 shows the largest additional cost incurred by GRASP in comparison to those obtained by the optimal method for the three considered networks and cases.

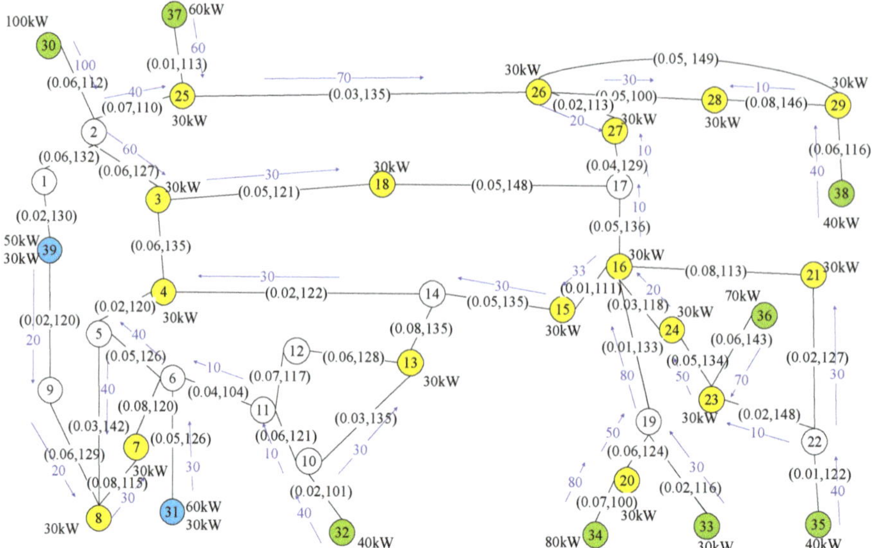

Figure 13. Path selection and source-load matching performed by GRASP on the IEEE39 bus.

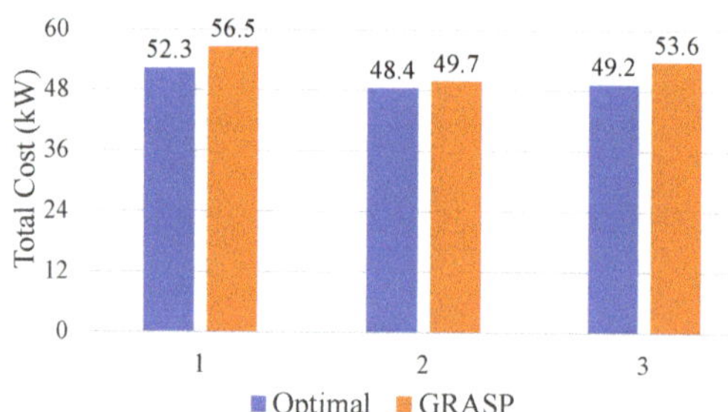

Figure 14. Comparison of power costs for the optimal solution and GRASP on the IEEE39 bus system.

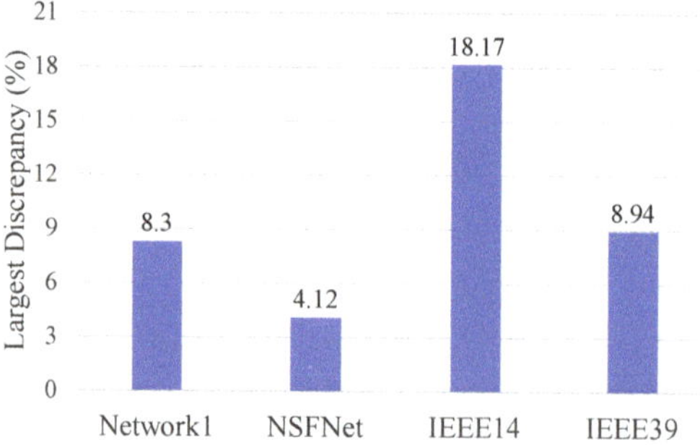

Figure 15. Summary of the largest additional costs required by GRASP for the different considered distribution networks.

3.2. An Example of a Practical Application of GRASP

We used low-frequency power data collected from six houses from the Reference Energy Disaggregation Data Set (REDD) [37]. The data of each house consisted of power consumption information measured from different appliances, such as a refrigerator, lighting, microwave, etc.

The data contained power consumption from real homes, for the whole house as well as for each circuit in the house (labeled by the main type of appliance on that circuit).

We used the power consumption profile of six different houses as loads and five different energy sources. Figure 16 shows the power profiles for 24 h of the six loads. The energy resources were three photovoltaic (solar) generators [38] and two wind turbine generators [39]. We used the same power profile for the solar and wind sources. Figure 17 shows the profile of the solar and wind generators.

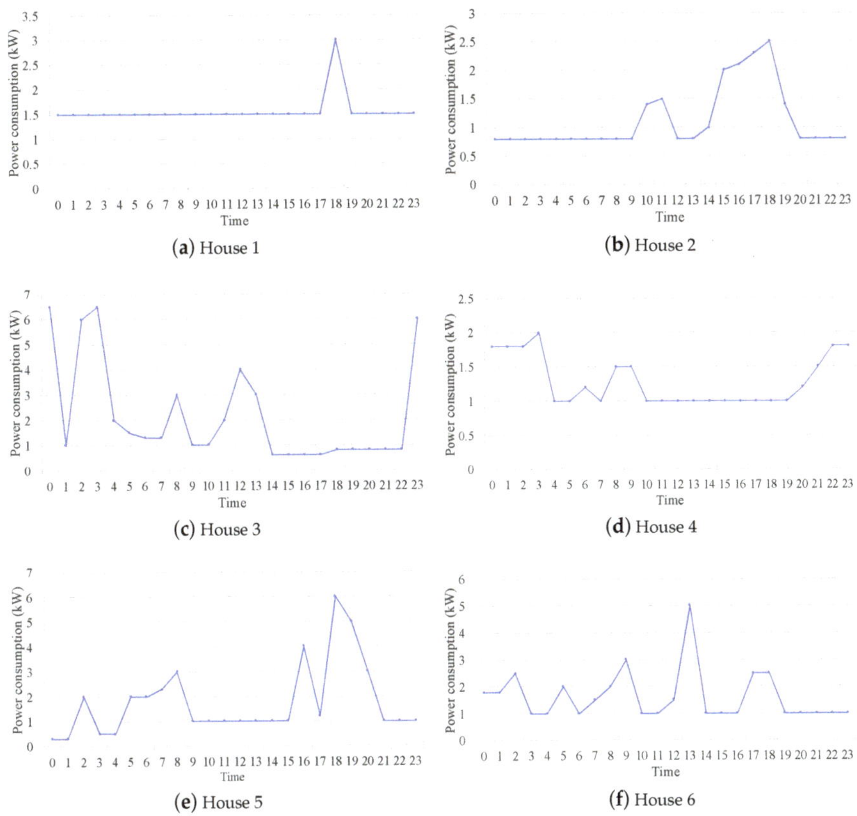

Figure 16. Practical case with the consumption of six loads in IEEE14 bus.

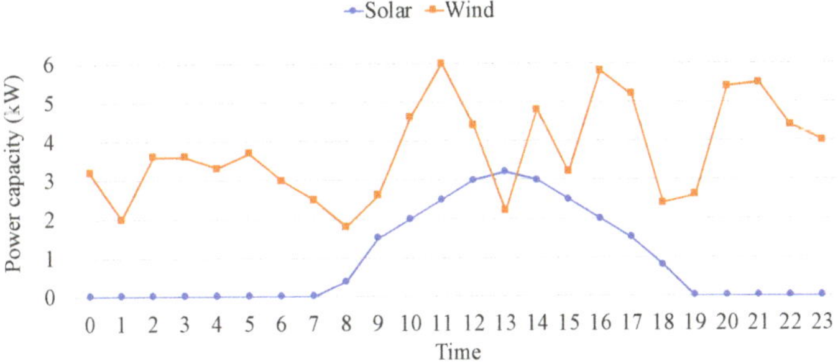

Figure 17. Generation of power in practical sources for solar and wind turbine.

We used the IEEE 14 testbus to interconnect the sources and loads. In this practical example, we used one generator of constant power (power line) as node 1, two identical Solar generators as nodes 2 and 3, and two identical wind-turbine generators, as nodes 6 and 8. The power capacity of the constant power generator is 10 kW. The loads in this testbus were Nodes 12, 13, 14, 11, 5, and 4 as houses 1–6, respectively.

We performed power routing for this example, and the summary of per-hour distribution provides the results Figure 18 shows. These results show that the largest cost discrepancy is 0.1 kW at 6 p.m. Figure 19 shows the routing of power at 6 p.m. for the optimal method and GRASP.

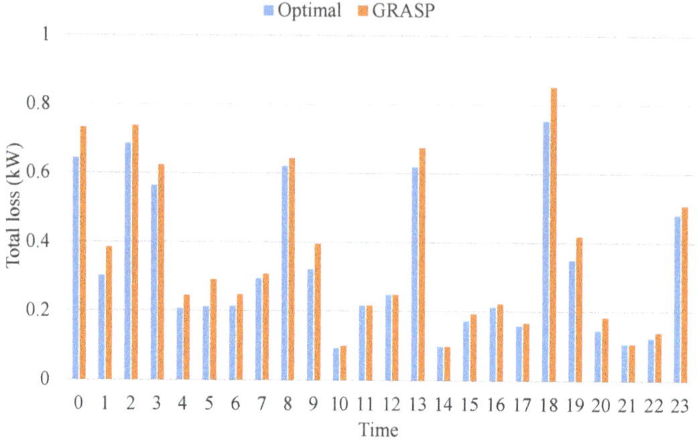

Figure 18. Summary of the costs difference required by GRASP and optimal method in a practical example.

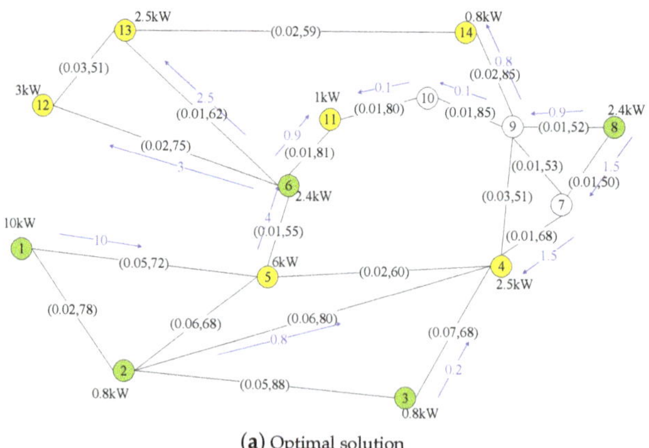

(**a**) Optimal solution

Figure 19. *Cont.*

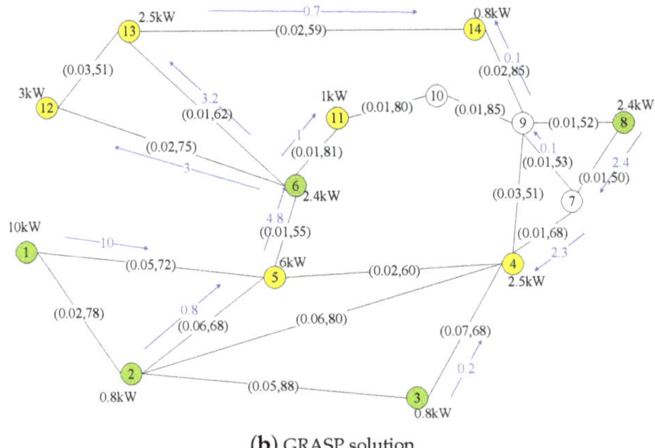

(b) GRASP solution

Figure 19. Routing of power at 6 pm on IEEE14 testbus by (a) the optimal method and (b) GRASP on the practical example.

4. Conclusions

In a source-dense digital microgrid, the distribution of power consists of matching loads to the power sources that can satisfy their power request, and of the selection of the lowest-cost paths interconnecting them. Such paths transfer the requested amounts of power but with minimum total costs. We considered the costs incurred by a function of the power transferred through the link. To realize power routing, we first modelled the problem as an integer linear programming approach. While this method provides an optimal solution, its complexity is NP-hard. Therefore, the application of this optimal method on a digital microgrid with a large number of sources is prohibitive. As a solution, we have proposed the greedy smallest-cost-rate path first algorithm, or GRASP in short, to assign sources to loads and obtain the smallest cost routes between sources and loads. We tested GRASP on four different network topologies, including IEEE test buses. We compared the total cost obtained by the optimal method and GRASP. The results show that GRASP achieves comparable costs as does the optimal solution, with a discrepancy of up to 18 % on the considered scenarios. We also present a practical application of GRASP where we use the power consumption profile of six houses and the actual power generation profile of five sources on IEEE14 testbus and show the differences between the optimal method and GRASP. We showed that the largest discrepancy cost in this example is 0.1 kW. Our analysis also shows that GRASP achieves these additional costs in exchange for a computation complexity smaller than that of the optimal method.

Author Contributions: conceptualization, Z.J., V.S., and R.R.-C.; methodology, V.S. and Z.J.; investigation, Z.J., H.G. and R. R.-C.; writing, review and editing, Z.J., H.G., A.M., and R.R.-C.; project administration, R.R.-C.

Funding: This research was supported in part by the NSF Award CNS 1641033.

Conflicts of Interest: The authors declare no conflict of interest.

References

1. Farhangi, H. The path of the smart grid. *IEEE Power Energy Mag.* **2010**, *8*, 18–28. [CrossRef]
2. Office of Electricity Delivery and Energy Reliability. Grid Modernization and the Smart Grid. 2010. Available online: https://www.energy.gov/oe/activities/technology-development/grid-modernization-and-smart-grid (accessed on 20 April 2019).
3. Gungor, V.C.; Lu, B.; Hancke, G.P. Opportunities and challenges of wireless sensor networks in smart grid. *IEEE Trans. Ind. Electron.* **2010**, *57*, 3557–3564. [CrossRef]

4. Bu, S.; Yu, F.R.; Cai, Y.; Liu, X.P. When the smart grid meets energy-efficient communications: Green wireless cellular networks powered by the smart grid. *IEEE Trans. Wirel. Commun.* **2012**, *11*, 3014–3024. [CrossRef]
5. Feisst, C.; Schlesinger, D.; Frye, W. Smart grid: The role of electricity infrastructure in reducing greenhouse gas emissions. *Cisco Internet Bus. Solut. Group* **2008**. Available online: https://www.cisco.com/c/dam/en_us/about/ac79/docs/Smart_Grid_FINAL.pdf (accessed on 10 April 2019).
6. Yu, F.R.; Zhang, P.; Xiao, W.; Choudhury, P. Communication systems for grid integration of renewable energy resources. *IEEE Netw.* **2011**, *25*, 22–29. [CrossRef]
7. Saleh, M.S.; Althaibani, A.; Esa, Y.; Mhandi, Y.; Mohamed, A.A. Impact of clustering microgrids on their stability and resilience during blackouts. In Proceedings of the 2015 International Conference on Smart Grid and Clean Energy Technologies (ICSGCE), Offenburg, Germany, 20–23 October 2015; pp. 195–200.
8. Saleh, M.; Esa, Y.; Mohamed, A.A. Communication-Based Control for DC Microgrids. *IEEE Trans. Smart Grid* **2019**, *10*, 2180–2195. [CrossRef]
9. Ton, D.T.; Smith, M.A. The US department of energy's microgrid initiative. *Electricity J.* **2012**, *25*, 84–94. [CrossRef]
10. Matsumoto, Y.; Yanabu, S. A vision of an electric power architecture for the next generation. *Electrical Eng. Jpn.* **2005**, *150*, 18–25. [CrossRef]
11. Abe, R.; Taoka, H.; McQuilkin, D. Digital grid: Communicative electrical grids of the future. *IEEE Trans. Smart Grid* **2011**, *2*, 399–410. [CrossRef]
12. Hikihara, T.; Tashiro, K.; Kitamori, Y.; Takahashi, R. Power Packetization and Routing for Smart Management of Electricity. In Proceedings of the 10th International Energy Conversion Engineering Conference, Atlanta, GA, USA, 30 July–1 August 2012; pp. 1–6, doi:10.2514/6.2012-3732. [CrossRef]
13. Tashiro, K.; Takahashi, R.; Hikihara, T. Feasibility of power packet dispatching at in-home DC distribution network. In Proceedings of the 2012 IEEE Third International Conference on Smart Grid Communications (SmartGridComm), Tainan, Taiwan, 5–8 November 2012; pp. 401–405.
14. Xu, Y.; Rojas-Cessa, R.; Grebel, H. Allocation of discrete energy on a cloud-computing datacenter using a digital power grid. In Proceedings of the 2012 IEEE International Conference on Green Computing and Communications, Besancon, France, 20–23 November 2012; pp. 615–618.
15. Gelenbe, E. Energy packet networks: Adaptive energy management for the cloud. In Proceedings of the ACM 2nd International Workshop on Cloud Computing Platforms, Bern, Switzerland, 10 April 2012; pp. 1–5.
16. Girbau-Llistuella, F.; Rodriguez-Bernuz, J.; Prieto-Araujo, E.; Sumper, A. Experimental validation of a single phase intelligent power router. In Proceedings of the IEEE PES Innovative Smart Grid Technologies, Europe, Istanbul, Turkey, 12–15 October 2014; pp. 1–6.
17. Matsuura, K.; Taoka, H.; Kato, R.; Abe, R. Digital grid in low-voltage distribution system. In Proceedings of the 2015 IEEE Power & Energy Society General Meeting, Denver, CO, USA, 26–30 July 2015; pp. 1–5.
18. Abe, R. Digital grid: Full of renew able energy in the future. In Proceedings of the 2016 IEEE International Conference on Consumer Electronics-Taiwan (ICCE-TW), Nantou, Taiwan, 27–29 May 2016; pp. 1–2.
19. Nakano, H.; Okabe, Y. Toward implementing power packet networks. In Proceedings of the 2017 IEEE International Conference on Consumer Electronics-Taiwan (ICCE-TW), Taipei, Taiwan, 12–14 June 2017; pp. 199–200.
20. Saitoh, H.; Toyoda, J. A new concept of electric power network for the effective transportation of small power of dispersed generation plants. *IEEJ Trans. Power Energy* **1995**, *115*, 568–575. [CrossRef]
21. Medida, S.; Sreekumar, N.; Prasad, K.V. SCADA-EMS on the Internet. In Proceedings of the EMPD'98, 1998 International Conference on Energy Management and Power Delivery (Cat. No. 98EX137), Singapore, 5 March 1998; Volume 2, pp. 656–660.
22. Khatib, A.R.; Dong, Z.; Qiu, B.; Liu, Y. Thoughts on future Internet based power system information network architecture. In Proceedings of the 2000 Power Engineering Society Summer Meeting (Cat. No. 00CH37134), Seattle, WA, USA, 16–20 July 2000; Volume 1, pp. 155–160.
23. Reifenhaeuser, B.; Ebbes, A. Digitalized distribution system for the power supply. Internet of the energy; Digitalisiertes Verteilungsnetz fuer die Stromversorgung. Internet der Energie. *EW. Das Magazin fuer die Energie Wirtschaft* 2013, *112*. Available online: https://www.osti.gov/etdeweb/biblio/22110540 (accessed on 10 July 2019).

24. Rojas-Cessa, R.; Xu, Y.; Grebel, H. Management of a smart grid with controlled-delivery of discrete levels of energy. In Proceedings of the 2013 IEEE Electrical Power & Energy Conference, Halifax, NS, Canada, 21–23 August 2013; pp. 1–6.
25. Rojas-Cessa, R.; Sahasrabudhe, V.; Miglio, E.; Balineni, D.; Kurylo, J.; Grebel, H. Testbed evaluations of a controlled-delivery power grid. In Proceedings of the 2014 IEEE International Conference on Smart Grid Communications (SmartGridComm), Venice, Italy, 3–6 November 2014; pp. 206–211.
26. Shibano, K.; Kontani, R.; Hirai, H.; Hasegawa, M.; Aihara, K.; Taoka, H.; McQuilkin, D.; Abe, R. A linear programming formulation for routing asynchronous power systems of the Digital Grid. *Eur. Phys. J. Spec. Top.* **2014**, *223*, 2611–2620. [CrossRef]
27. Murty, K.G. *Linear Programming*; Springer: New York, NY, USA 1983.
28. Garfinkel, R.S. *Integer Programming*; Technical Report; Wiley: Hoboken, NJ, USA, 1972.
29. Dijkstra, E.W. A note on two problems in connexion with graphs. *Numerische Math.* **1959**, *1*, 269–271. [CrossRef]
30. Rojas-Cessa, R.; Wong, C.K.; Jiang, Z.; Shah, H.; Grebel, H.; Mohamed, A. An Energy Packet Switch for Digital Power Grids. In Proceedings of the 11th International Conference on Internet of Things (iThings 2018), Halifax, NS, Canada, 30 July–3 August 2018; pp. 1–8.
31. Papadimitriou, C.H. On the complexity of integer programming. *J. ACM (JACM)* **1981**, *28*, 765–768. [CrossRef]
32. Barbehenn, M. A note on the complexity of Dijkstra's algorithm for graphs with weighted vertices. *IEEE Trans. Comput.* **1998**, *47*, 263. [CrossRef]
33. Frazer, K. *NSFNET: A Partnership for High-Speed Networking*; Merit Network: Ann Arbor, MI, USA, 1995.
34. Washington University. IEEE14 Test Bus. Available online: https://labs.ece.uw.edu/pstca/pf14/pg_tca14bus.htm (accessed on 19 May 2019).
35. Athay, T.; Podmore, R.; Virmani, S. A practical method for the direct analysis of transient stability. *IEEE Trans. Power Apparatus Syst.* **1979**, *PAS-98*, pp. 573–584. [CrossRef]
36. ICSEG. IEEE 39 Test Bus. Available online: https://icseg.iti.illinois.edu/ieee-39-bus-system/ (accessed on 18 May 2019).
37. Kolter, J.Z.; Johnson, M.J. REDD: A public data set for energy disaggregation research. In Proceedings of the Workshop on Data Mining Applications in Sustainability (SIGKDD), San Diego, CA, USA, 21 August 2011; Volume 25, pp. 59–62.
38. Parsons, B.; Hummon, M.; Cochran, J.; Stoltenberg, B.; Batra, P.; Mehta, B.; Patel, D. *Variability of Power from Large-Scale Solar Photovoltaic Scenarios in the State of Gujarat*; Technical Report; National Renewable Energy Lab. (NREL): Golden, CO, USA, 2014.
39. Nguyen, T.V. Integration of compressed air energy storage with wind turbine to provide energy source for combustion turbine generator. In Proceedings of the IEEE PES Innovative Smart Grid Technologies, Europe, Istanbul, Turkey, 12–15 October 2014; pp. 1–5.

Article

An Effective Passive Islanding Detection Algorithm for Distributed Generations

Arash Abyaz [1], Habib Panahi [1], Reza Zamani [2], Hassan Haes Alhelou [3], Pierluigi Siano [4,*], Miadreza Shafie-khah [5] and Mimmo Parente [4]

1 School of Electrical and Computer Engineering, University of Tehran, Tehran 395515, Iran
2 Faculty of Electrical and Computer Engineering, Tarbiat Modares University, Tehran 115111, Iran
3 Department of Electrical Power Engineering, Faculty of Mechanical and Electrical Engineering, Tishreen University, Lattakia 2230, Syria
4 Department of Management & Innovation Systems, University of Salerno, 84084 Salerno, Italy
5 School of Technology and Innovations, University of Vaasa, 65200 Vaasa, Finland
* Correspondence: psiano@unisa.it; Tel.: +39-320-4646-454

Received: 1 July 2019; Accepted: 12 August 2019; Published: 16 August 2019

Abstract: Different issues will be raised and highlighted by emerging distributed generations (DGs) into modern power systems in which the islanding detection is the most important. In the islanding situation, a part of the system which consists of at least one DG, passive grid, and local load, becomes fully separated from the main grid. Several detection methods of islanding have been proposed in recent researches based on measured electrical parameters of the system. However, islanding detection based on local measurements suffers from the non-detection zone (NDZ) and undesirable detection during grid-connected events. This paper proposes a passive islanding detection algorithm for all types of DGs by appropriate combining the measured frequency, voltage, current, and phase angle and their rate of changes at the point of common coupling (PCC). The proposed algorithm detects the islanding situation, even with the exact zero power mismatches. Proposed algorithm discriminates between the islanding situation and non-islanding disturbances, such as short circuit faults, capacitor faults, and load switching in a proper time and without mal-operation. In addition, the performance of the proposed algorithm has been evaluated under different scenarios by performing the algorithm on the IEEE 13-bus distribution system.

Keywords: Islanding detection; distributed generation; modern power system; anti-islanding; protection

1. Introduction

Modern power systems are facing high complexity and sensitivity. One of the aspects of these grids and also future power system is the high penetration of distribution generations (DGs) into the network, which raises several issues. Islanding detection is a serious problem that emerges in this environment. In the islanding situation, one or more of the DGs and a portion of the network become separated from the main grid and continue to energize the islanded network. Due to several concerns, such as safety hazard, personal safety, and out-of-phase reclosing, islanding detection has vital importance. The islanding should be detected less than two seconds, according to the IEEE 1547 standard [1] and also with the respect of the re-closer operation. The re-closer constant time usually is less than 500 ms, hence the time of detection is a critical issue to discriminate islanding before the operation of the re-closer. Furthermore, some grid codes [2] imply that the DG should have the capability to continue to operate in the autonomous islanded mode. Therefore, the islanding is detected to the purpose of disconnecting the DG or preparing the DG to operate in the new control mode after the occurrence of the islanding.

In general, islanding detection methods can be classified into remote and local techniques. Remote techniques detect the islanding situation using communication links between the DG and the point of common coupling (PCC), which connects the network to the upstream grid [3–6]. In Reference [5], islanding can be detected using PMU measurements. Power line signaling utilized in Reference [6] to detect the islanding by transmitting and an insignificant signal from the DG bus and receiving it at the circuit breaker point. Remote methods detect the islanding with a negligible or even zero none detection zone (NDZ). In addition, these techniques are fast in the detection of the islanding situation and discriminating islanding from grid-connected disturbances. However, the main drawbacks of the remote methods are their high cost, risk of loss of communication links, and requiring to the backup protection problem.

Local methods can be divided into active and passive ones. A disturbance is added to the system deliberately to detect the islanding situation in the active methods. Consequently, active methods can be designed based on the type of DG. Hence, the active methods are used for specified types of DGs. In addition, in some active methods, the disturbance signal is not added to the system by means of the DG. For instance, inserted capacitor or condenser to intentionally disturbing the equality power mismatches to detect the islanding situation [7,8]. Some of the active methods for inverter-based generators are Sandia frequency shift [9], current harmonic injection [10], active frequency drift [11], voltage drifting technique [12], and positive feedback method [13]. Capability issue of current injection method is investigated in Reference [14] and a solution to solve this problem is also introduced based on considering only certain high frequency components as injected current. Likewise, in References [15–17], islanding detection methods for synchronous based generators are presented. In Reference [15], two positive feedback for active and reactive power control loops are presented to make the system unstable in the islanding situation. Afterwards, in Reference [17], new active and passive power control loops proposed to simultaneously improve the ride-through capability of the synchronous generator, as well as detecting islanding-situation. Integral controllers are added to the governor, and excitation system of the synchronous generator are added in Reference [16] to make the system marginally unstable in the islanding situation. In Reference [18], two modified active and reactive power control loops of synchronous DG is presented to make the system marginally unstable in islanding situation. Moreover, a signal processing technique is used to detect a feature which indicates the islanding conditions. A method to control the active and reactive power outputs of inverter-based DGs using two probabilistic phasing neural network controllers has been presented in Reference [19]. Active techniques can interfere with other devices which inject the disturbance to the system and also non-linear loads. Also, power quality degradation is another drawback of the active islanding methods.

Passive methods are based on monitoring the electrical quantities of the system. In this manner, the islanding is detected by violating the presets thresholds. Over/under voltage or frequency relays are the most common passive loss of mains detection [20]. Vector surge relays [21] are another type of passive islanding detection technique. Rate of change of frequency (ROCOF) is another common islanding detection method [22] which has large NDZ. In addition, setting the threshold of the ROCOF is a direct compromise between reducing the NDZ and maloperation in non-islanding events. Also, the voltage measurement-based methods, such as the rate of change of voltage (ROCOV) [8] have an unsuccessful operation to distinguish islanding condition, especially in the case of reactive power imbalance. Afterwards, the rate of change of phase angle difference (ROCPAD) is presented in Reference [23]. The maloperation in short circuit fault is the main drawback of these methods. Signal processing techniques are other passive methods that have been introduced recently [24–26]. The wavelet transform is used in References [25,26] to detect the islanding and minimize the NDZ. A passive method for inverter-based DGs introduced in Reference [27] by combining the help of close loop frequency control and a high frequency impedance detection method. In Reference [28], learning methods have been developed to extract the features which illustrate the difference between islanding and grid connected conditions. At first, different features are analyzed using signal processing methods and then these features applied to a deep learning-based algorithm to classified islanding and grid connected disturbances. A voltage index has been used in Reference [29] to detect the islanding condition for large power imbalances, also for small power imbalances, the line current measured to disconnect some loads, due to transfer small imbalances to large power mismatches.

This paper proposes a passive islanding detection technique to eliminate the NDZ and improve the detection time. The measured values of the frequency, voltage, and current are used in the proposed technique to discriminate between islanding and non-islanding conditions. Also, a method to accurately calculate the phase angle difference between voltage and current is proposed in this paper for real-time applications. In addition, the proposed islanding detection technique can be used for all types of the DGs and without depending on the load type of the system. Performance of the proposed method can be retained when multiple DGs are connected to the islanded network. The maximum islanding detection time using the proposed islanding technique is 6 ms even though the power mismatch is zero. The detection time is reduced by increasing the power imbalances based on the proposed technique. The main contributions of this paper can be summarized as follows:

- The detection time of islanding conditions is improved significantly using the proposed novel islanding detection method;
- Mal-operation in grid connected disturbances, such as short circuit faults are well-prevented using the proposed method;
- Since the proposed method is simple and is based on common protection relays, and it can be easily implemented practically;
- The proposed method can be applied for all types of DGs without degrading the power quality of the system.

2. Proposed Algorithm

Many methods have been proposed for islanding detection. The main concept of most islanding detection methods is that some system parameters (like the voltage, frequency, etc.) change greatly in islanding mode, while they do not change significantly when the distribution system is grid connected. In some cases, existing methods do not recognize islanding mode and cause mal-operation. In addition, these methods may have maloperation in power system disturbances like short circuit faults. This paper, by combining different methods that are described below, provides a new algorithm that has a better operation for islanding detection.

One method for generator's islanding detection is the use of under frequency (UF) and over frequency (OF) relays (Figure 1). When islanding mode happens due to the power mismatch between load and generation, generator's speed changes according to Equation (1) [30].

$$2 \times \mathrm{H} \times \frac{d\omega}{dt} = P_m - P_e = P_{mismatch}, \tag{1}$$

where:

H	Inertia constant (MW-s/MVA)
ω	Synchronous speed (p.u.)
P_m	Mechanical power (p.u.)
P_e	Electrical real power (p.u.)

Frequency depends on speed; therefore, when speed changes, the frequency will also change. If the generation is greater than demand, the frequency will increase, and it will decrease if the generation is less than load. The islanding mode will be detected if the frequency drops below or increases above a defined setting for a time greater than a defined delay.

In this paper, frequency is calculated using the PCC voltage. First, the frequency range is estimated using the zero-crossing method. Then, using the least squares error (LSE) method, it is calculated accurately.

Consider the main signal as follows:

$$V(t) = V_p Sin(\omega t + \theta), \tag{2}$$

where:

$V(t)$ Measured voltage (V)
ω Angular speed (rad/s)
V_P Peak of voltage (V)
θ Angle (rad)
t Time (s)

By sinusoidal expansion, Equation (2) becomes Equation (3).

$$V(t) = V_p Cos\theta Sin(\omega t) + V_p Sin\theta Cos(\omega t). \tag{3}$$

Equation (3) is expended by using Taylor's sinus and cosine expansions to obtain Equation (4).

$$\begin{aligned} V(t) = & Sin(\omega t)V_p Cos\theta + (\Delta\omega)tCos(\omega t)V_p Cos\theta + \\ & Cos(\omega t)V_p Sin\theta - (\Delta\omega)tSin(\omega t)V_p Sin\theta - \\ & \tfrac{t^2}{2}Sin(\omega t)(\Delta\omega)^2 Cos\theta - \tfrac{t^2}{2}Cos(\omega t)(\Delta\omega)^2 Sin\theta \end{aligned} \tag{4}$$

where $\Delta\omega$ is equal to change in angular speed. There are six unknowns in Equation (4), so at least seven equations are needed to calculate the unknowns by LSE method. In other words, at least seven samples are required to estimate the frequency. To calculate angular speed variations, the known variable and unknown variable of Equation (4) are separated and written as Equation (5). Finally, by adding frequency changes to the base frequency obtained from zero crossings, the frequency is calculated according to Equation (7).

$$V(t) = S_{11}X_1 + S_{12}X_2 + S_{13}X_3 + S_{14}X_4 + S_{15}X_5 + S_{16}X_6, \tag{5}$$

Where,

$$\begin{array}{lll} S_{11} = V_p Cos\theta & S_{12} = tV_p Cos\theta & S_{13} = V_p Sin\theta \\ S_{14} = -tV_p Sin\theta & S_{15} = -\frac{t^2}{2}Cos\theta & S_{16} = -\frac{t^2}{2}Sin\theta \end{array}$$

$$\begin{array}{lll} X_{11} = Sin(\omega t) & X_{12} = (\Delta\omega)Cos(\omega t) & X_{13} = Cos(\omega t) \\ X_{14} = (\Delta\omega)Sin(\omega t) & X_{15} = Sin(\omega t)(\Delta\omega)^2 & X_{16} = Cos(\omega t)(\Delta\omega)^2 \end{array}$$

$$\Delta\omega = \frac{X_2 + X_4}{X_1 + X_3} \tag{6}$$

$$f = f_{zero} + \frac{\Delta\omega}{2\pi f_0} \tag{7}$$

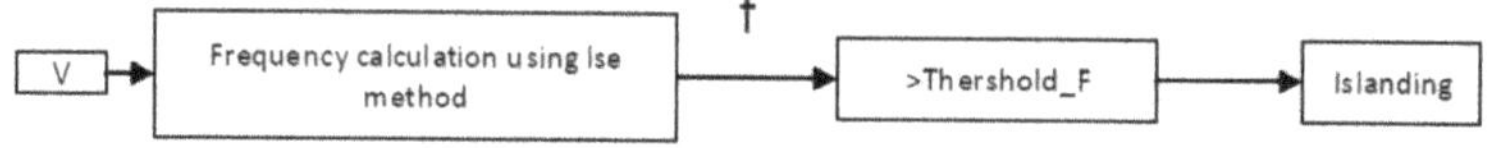

Figure 1. Islanding detection using under frequency (UF) and over frequency (OF) relays.

Rate of change of frequency (ROCOF) is an important technique for islanding detection. In low power mismatches, the frequency decreases slowly, and it takes a long time to exceed the relay set point. Therefore, islanding mode is detected faster using the ROCOF method. The ROCOF method is based on the theory that there is a mismatch between generation and load when islanding mode happens. Immediately after islanding, because of the power mismatch, the frequency will change dynamically, which neglecting the governor action can be approximated by Equation (8).

$$\frac{df}{dt} = -\left(\frac{P_L - P_G}{2H \times S_{GN}} \times f_r\right), \tag{8}$$

where:

f	**Frequency (Hz)**
P_L	Load in the island (MW)
P_G	Output of the distributed generator (MW)
S_{GN}	Distributed generator rating (MW)
H	Inertia constant of generating plant (MW-s/MVA)
f_r	Rated frequency (Hz)

ROCOF is determined by deriving of the frequency which is obtained by the LSE method. The ROCOF is measured to compare it with the set threshold limit. If its value exceeds the pre-specified threshold limit, islanding is detected. The block diagram of the ROCOF method is shown in Figure 2.

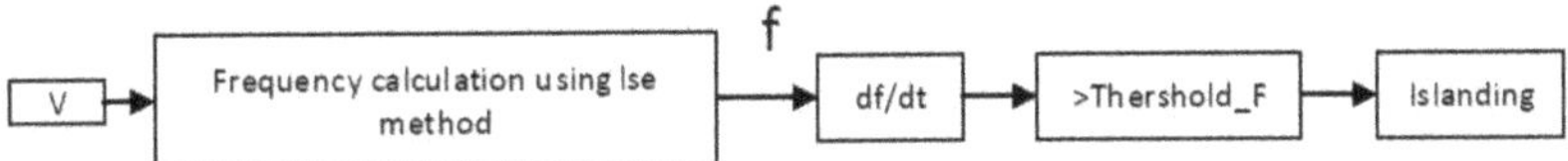

Figure 2. Islanding detection using rate of change of frequency (ROCOF) method.

Another method of islanding detection is ROCPAD. The fundamental of this method is based on the rate of change of phase angle difference at DG end. The process starts with measuring voltage and current signal at the DG end, and then from the phase angle information of the voltage and current signals, the ROCPAD is computed for islanding detection. The five-sample LSE method is used to find the current and voltage angle.

The voltage signal is sampled at $t = t_0$:

$$V(t_0) = V_p \mathrm{Sin}(\omega_0 t_0 + \theta_V), \tag{9}$$

where θ_v is the phase angle of the voltage signal. Equation (9) can be written as Equation (10):

$$V(t_0) = V_p \mathrm{Cos}\theta_V \mathrm{Sin}(\omega_0 t_0) + V_p \mathrm{Sin}\theta_V \mathrm{Cos}(\omega_0 t_0). \tag{10}$$

Consider Equation (10) as Equation (11):

$$V(t_0) = a_{01} X_1 + a_{02} X_2, \tag{11}$$

where,

$$a_{01} = \mathrm{Sin}\ (\omega_0 t_0) a_{02} = \mathrm{Cos}\ (\omega_0 t_0),$$

$$X_1 = V_p \mathrm{Cos}\theta_V X_2 = V_p \mathrm{Sin}\theta_V.$$

It is done for four other voltage samples, and five equations are obtained similar to Equation (11). The matrix form of these equations is given by Equation (12),

$$[A]_{5\times2} \cdot [X]_{2\times1} = [V]_{5\times1}. \tag{12}$$

The unknown values of matrix X can be calculated from Equation (13),

$$[X]_{2\times1} = [A]^{+}_{2\times5} \cdot [V]_{5\times1}, \tag{13}$$

$$[A]^{+} = \left[[A]^{T}.[A]\right]^{-1}.[A]^{T}. \quad (14)$$

Finally, voltage angle is calculated from Equation (15). It is also done for the current signal to get the current angle.

$$\theta = \tan^{-1}\left(\frac{X_2}{X_1}\right) \quad (15)$$

The ROCPAD can be calculated by deriving of the difference between the voltage and current angle. Islanding happened when its value exceeds a defined threshold. The block diagram of the ROCPAD method is shown in Figure 3.

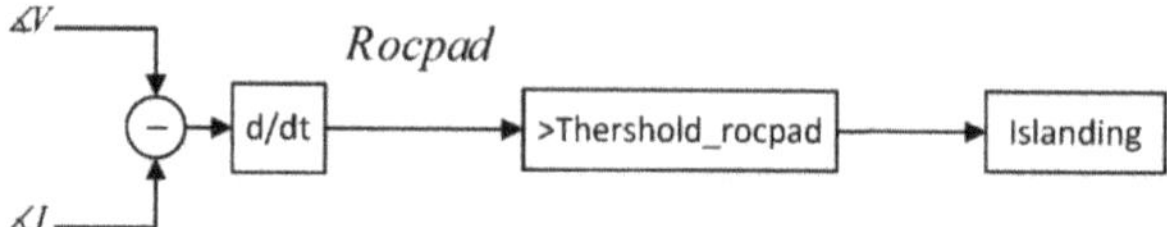

Figure 3. Islanding detection using rate of change of phase angle difference (ROCPAD) method.

The rate of change of voltage, which is known as ROCOV index is one of the islanding detection methods. Most of DGs are required to operate at unit power factor. Thus, the lack of active power will be possible. Change of voltage is a function of reactive power, and as a result of the change of reactive power, the voltage will be changed. Capacitor banks may be the only available source of reactive power in the islanded network with DG operating in unit power factor. The amount of reactive power generated by capacitor banks is a function of the voltage. When the voltage changes, as a result of islanding, the reactive power which capacitor banks generate will also change, and it will change the voltage further. Thus, when reactive power mismatch has a large value, this causes a change in the voltage, so ROCOV value exceeds a preset threshold and islanding mode is detected.

Different methods of Islanding detection are described above. Each of these methods has problems that cause mal-operation or maloperation. The operation of OF/UF relays depends on the frequency. If the mismatch between the generation and load in the islanded network is low, then the frequency will have small changes. Therefore, it may not exceed the threshold limit and leads to false detection. On the other hand, when short circuit faults happen in power systems, the frequency decreases, due to a sudden decrease in load. In this situation, this method may operate incorrectly.

ROCOF method measures the rate of change of frequency, and once the rate of change of frequency exceeds the pre-determined setting, a trip signal is initiated. Similar to OF/UF relays, when active power imbalance in the islanding system is low, the frequency changes slightly. Therefore, the ROCOF method may become ineffective. In addition, network transient events may also cause changes in system frequency, resulting in the incorrect operation of the ROCOF method.

The reactive power imbalance is the detection criterion of the ROCOV method. If a load with a low power factor is disconnected from the network in a normal situation, the reactive power imbalance will have a large value. Therefore, the voltage changes and the ROCOV method may operate incorrectly. On the other hand, when reactive power imbalance in the islanding mode has small variation, the ROCOV value does not sense the threshold and cannot detect the islanding mode.

The ROCPAD method was proposed to solve the maloperation of the ROCOF, and OF/UF relays in the low power mismatch. This method even detects islanding mode with 0% power mismatch. However, the ROCPAD method may have mal-operation during a short circuit fault.

This paper, by combining ROCOF, ROCPAD, ROCOV, and OF/UF relays, presents an algorithm that solves the problems of previous methods. All previous methods can detect islanding mode. However, a more secure operation is achieved using the proposed algorithm. The proposed islanding detection algorithm is shown in Figure 4.

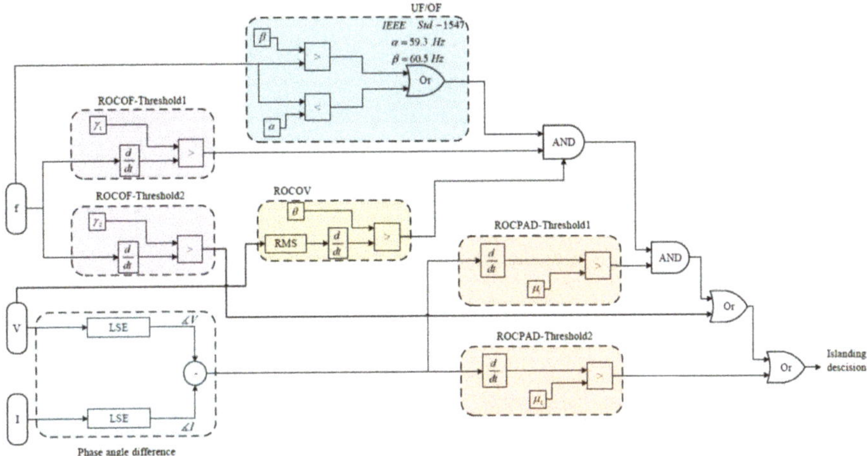

Figure 4. Proposed islanding detection algorithm.

In the proposed algorithm, the combination of ROCOF, ROCPAD, ROCOV and OF/UF relays is used to detect islanding mode fast. The operation time is an important parameter in protection algorithms. There is a direct relationship between the time of detection and the threshold value in classic methods. In order to fast operation, the ROCOV, ROCOF, OF/UF and ROCPAD methods with suitable setting are combined in the proposed method. As the operation speed of the classic algorithm increases, the probability of mal-operation will also increase. The proposed method can provide security and fast operation simultaneously.

The proposed algorithm can tackle the problems of classic algorithms by a logical combination of their output. The ROCPAD relay may have mal-operation in power system short circuit faults. While the mal-operation problem is solved by raising the threshold of this relay, the operation time increases. In the proposed method, two ROCPAD relays with different threshold are used for fast and secure operation. The first ROCPAD relay (ROCPAD-threshold1) has a lower threshold and is faster.

As mentioned above, the ROCOF, ROCPAD and OF/UF relays may have mal operation in power system short circuits. In short circuit faults, the voltage suddenly drops and reaches to a new value after a short time, but the voltage drops slowly in islanding mode. Using this rule, short circuit faults are separated from islanding modes. The ROCOV relay which measures the rate of change of voltage can separate short circuit faults from islanding mode. Therefore, the output of the ROCOV, ROCOF, ROCPAD and OF/UF relays are connected together by a logical AND gate to limit their operation in short circuit faults.

The second ROCPAD relay (ROCPAD-threshold2) has a higher threshold. Hence, it does not recognize short circuit faults as islanding mode. Its trip signal is connected to the output with a logical OR gate. If the first ROCPAD relay is limited, islanding will be detected through the second ROCPAD relay.

There is a direct relationship between the time of detection and the threshold of ROCOF relay. By reducing the threshold, as the islanding detection speed is increased, the probability of wrong detection increases. Therefore, Similar to ROCPAD relay, two ROCOF relays with different threshold are used in the proposed method. The first ROCOF relay (ROCOF-threshold1) has a lower threshold and operate faster. Noises and the transient faults change the frequency quickly. In this situation, the ROCOF value may exceed its threshold, while the frequency does not change significantly. In order to block the operation of the first ROCOF relay under transient faults and noises, the trip signal of it and OF/UF relay combined with a logical AND gate. The second ROCOF relay (ROCOF-threshold2) with higher threshold operates in parallel with second ROCPAD relay to enhance security.

The operation of the proposed algorithm in three general network modes is as follows:

- When islanding mode happens with high power mismatch, the frequency and voltage change significantly. As a result, the ROCOF, ROCOV and OF/UF relays detect islanding mode, and the output of their AND gate becomes one. On the other hand, the voltage and current phase angle difference begin to increase in islanding network and lead to the output of ROCPAD relay become one. Finally, the output of the second AND gate becomes one and islanding will be detected. In this case, the second ROCPAD and ROCOF relays will detect islanding mode with a time delay after islanding detection by a combination of the first ROCPAD. First, ROCOF, ROCOV and OF/UF relays. Fast operation in high power mismatch is essential, due to high power mismatch causes the generator to become unstable and damages to network equipment. The proposed algorithm by a combination of ROCPAD, ROCOF, ROCOV, and OF/UF relays with low threshold provides fast operation in high power mismatch.
- When islanding mode happens with low power mismatch, ROCOF, ROCOV, and OF/UF relays may not detect islanding mode. As a result, the output of their AND gate becomes zero. The voltage and current phase angle difference even in low power mismatch increases. Thus, the ROCPAD value exceeds the threshold limits, and the output of ROCPAD relay becomes one. The output of AND gate related to the output of ROCPAD relay and output of first and gate becomes zero. Therefore, the ROCOF, ROCOV and OF/UF relays limit the operation of the first ROCPAD relay. In this situation, the islanding mode is detected by the second ROCPAD relay with a time delay. In islanding mode with low power mismatch, the voltage and frequency change slightly in the islanded grid. Thus, it is last a longer time for the generator to become unstable, and the generator can be separated from the islanded grid with more time delay.
- When short circuit faults happen, ROCPAD, ROCOF, and OF/UF relays may have mal-operation. Their operation is limited by the ROCOV relay, and the output of second AND gate becomes zero. The second ROCPAD and ROCOF relays, due to their higher threshold do not have mal-operation during short circuit faults. Finally, the output of the trip signal combination becomes zero, and the proposed algorithm does not operate in short circuit faults properly.

3. Simulation and Results

In this paper, the voltage and current measured at the PCC by using capacitor voltage transformer (CVT) and current transformer (CT). Moreover, the least square error algorithm with five samples and seven samples, respectively, has been used to find the voltage and current phase angle and frequency of power system. The sampling rate is 1000 Hz. IEEE 13-bus distribution test feeder, shown in Figure 5, is used to simulate the proposed algorithm. The power system frequency is 60 Hz. Two gas generators are connected to the bus 12. The rated power of each generator is 2 megawatts. At bus 10, there is a 1 megawatt solar power plant connected to the network by an inverter. More details about generators are described in the Appendix A. There is also a 1.2 MVAr compensation capacitor on the network connected to the bus 13.

There is a fuse in the transmission line between the bus 6 and bus 7. The islanding mode happens due to burning the fuse. In all the cases in this paper, the islanding mode has happened, the transmission line between the bus 6 and bus 7 is disconnected. Simulations are performed using MATLAB and PSCAD software. In this paper, IEEE 13-bus distribution system is simulated by PSCAD software. The voltage and current obtained from the PSCAD at PCC location are used for islanding detection. The frequency estimation and the voltage and current phase angle extraction algorithms are developed in MATLAB software. Moreover, the proposed islanding detection method is simulated in MATLAB software. Specifically, the PSCAD is used to simulate the studied system and to generating the data considering different simulation scenarios as explained in the paper, while the MATLAB is used to implement the proposed islanding detection algorithm. In this case, the real power system behavior and the implementation of islanding detection are modeled the same to the reality. As a result, by the parallel execution of the PSCAD and MATLAB, the detection time and other implementation technical issues can be well-considered and determined.

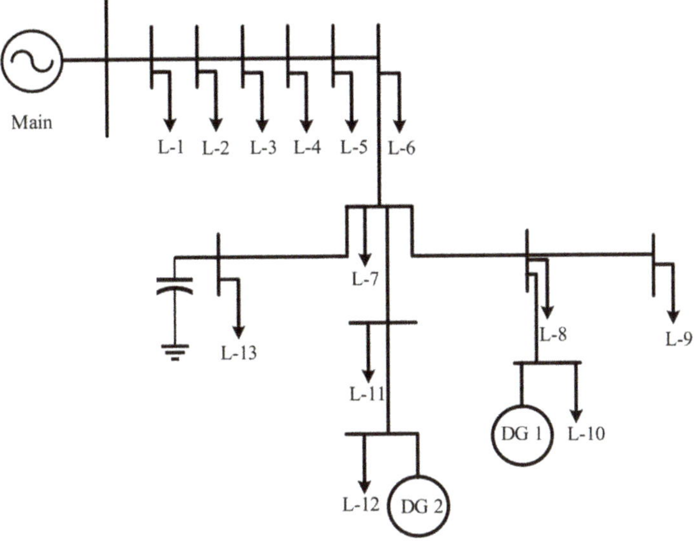

Figure 5. IEEE 13-bus distribution test feeder.

3.1. Islanding with High Power Mismatch

In the first case study, the power mismatch is very high. At t = 3, islanding mode happens with 50% power mismatch. The output of the ROCPAD-1, ROCPAD-2, ROCOF-1, ROCOF-2, ROCOV, and OF/UF relays is shown in Figures 6–9.

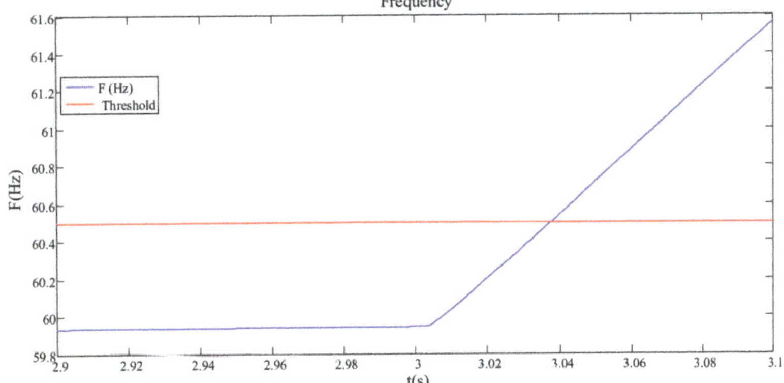

Figure 6. The output of OF/UF relay in islanding mode with high power mismatch.

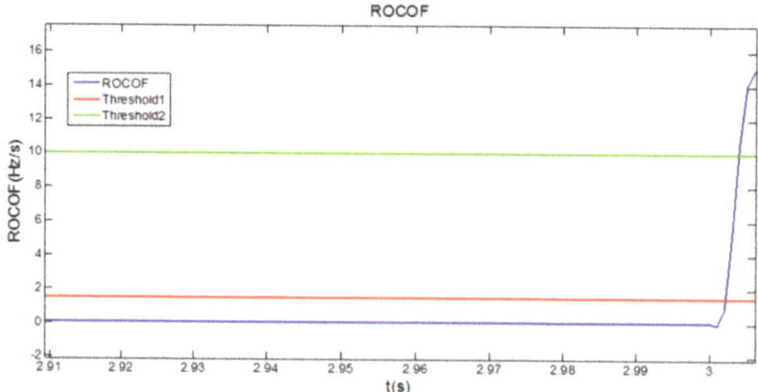

Figure 7. The output of ROCOF relay in islanding mode with high power mismatch.

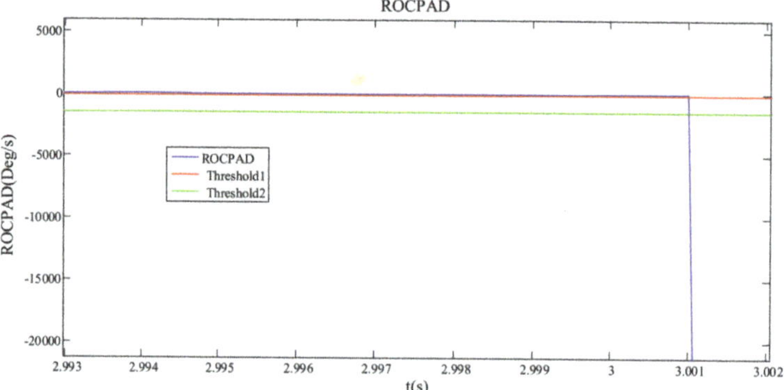

Figure 8. The output of ROCPAD relay in islanding mode with high power mismatch.

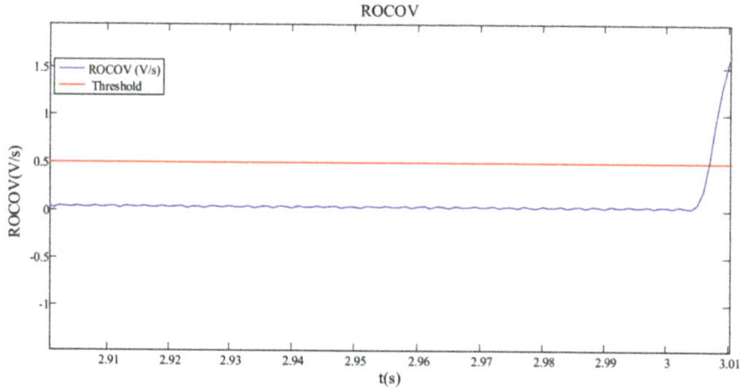

Figure 9. The output of ROCOV relay in islanding mode with high power mismatch.

As shown in Figure 6, after the islanding happens, the frequency begins to increase immediately and after a short time exceeds the threshold. Figure 7 shows that the ROCOF value increases rapidly and exceeds both thresholds, due to high power mismatch. The ROCPAD value is shown in Figure 8. The phase angle difference between the voltage and current begin to increase rapidly, and ROCPAD value exceeds both thresholds in a short time. Figure 9 shows that the ROCOV value increases rapidly and exceeds its threshold of less than 10 ms.

The trip signal of relays and the proposed algorithm is shown in Figure 10. It can be seen that islanding has been quickly detected by all relays and the proposed algorithm.

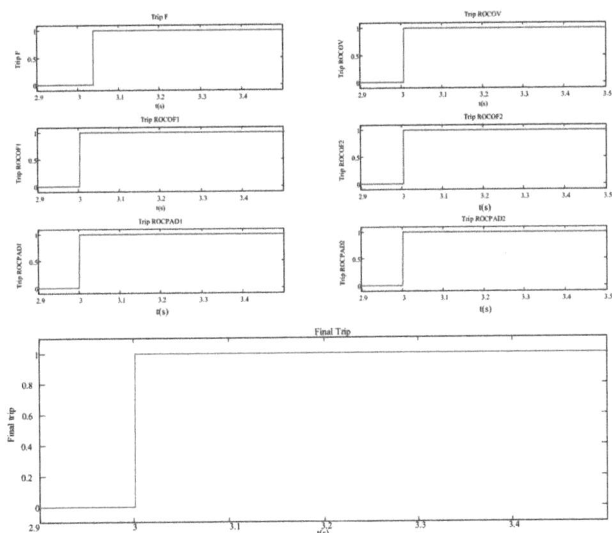

Figure 10. Trip signal of relays and the proposed algorithm in islanding mode with high power mismatch.

3.2. Short Circuit Fault

In the second case study, a short circuit fault occurs in the middle of the transmission line between bus 11 and bus 12 at t = 3. The short circuit resistance is equal to 1.5 Ω. The output of the ROCPAD-1, ROCPAD-2, ROCOF-1, ROCOF-2, ROCOV and OF/UF relays is shown in Figures 11–14.

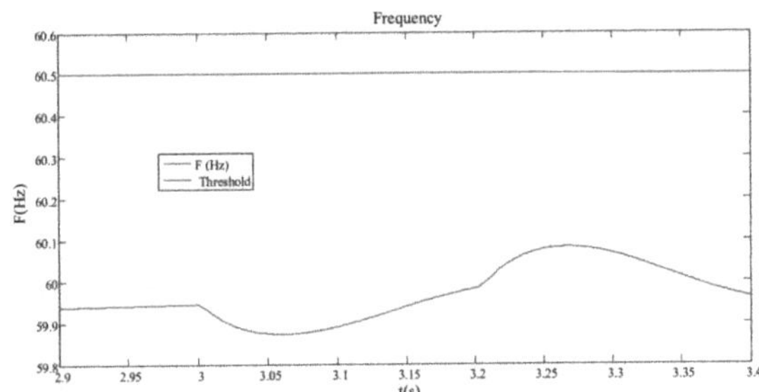

Figure 11. The output of OF/UF relay in short circuit fault.

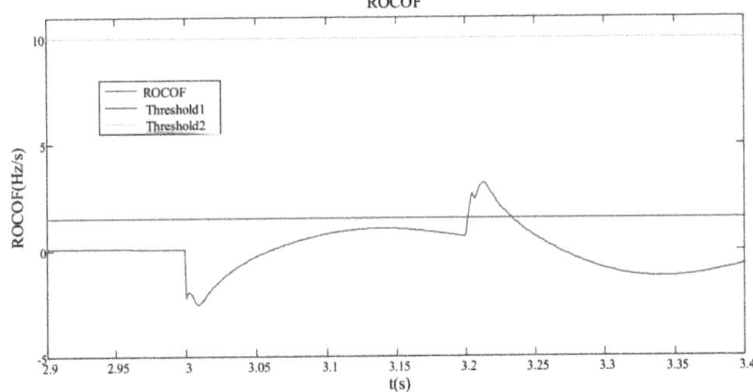

Figure 12. The output of ROCOF relay in short circuit fault.

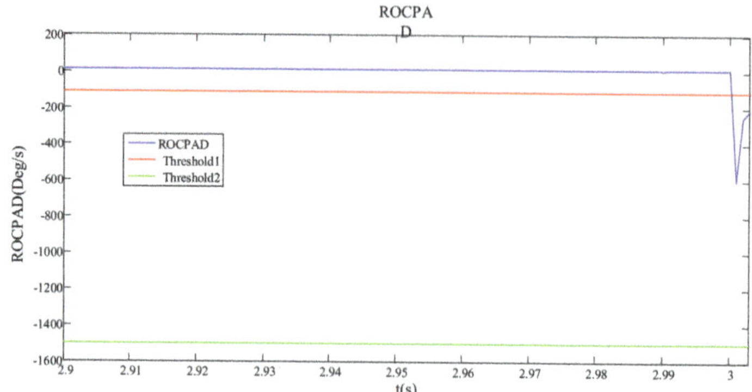

Figure 13. The output of ROCPAD relay in short circuit fault.

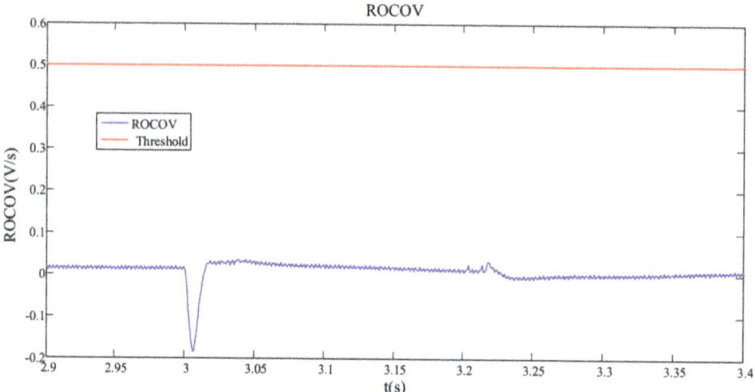

Figure 14. The output of ROCOV relay in short circuit fault.

As shown in Figure 11 in the short circuit fault, the frequency remains in an acceptable range. Therefore, the OF/UF relay does not work. Figures 12 and 13 show the ROCPAD and ROCOF value, respectively. It can be seen that the ROCPAD-threshold1 and ROCOF-threshold1 relays exceed their threshold. Thus, they operate incorrectly in short circuit faults. As mentioned above, after short circuit faults, the voltage drops immediately and reaches to a new value. On the other hand, the voltage drops with a lower slop in islanding modes. Thus, the ROCOV relay can separate islanding modes from non-islanding modes. As shown in Figure 14, the ROCOV value does not exceed the threshold and works correctly. Therefore, the output of the combination of these 4 relays with AND gate is zero. In this situation, if the ROCPAD and ROCOF are used, the generator is wrongly separated from the power grid.

Raising the ROCOF and ROCPAD relays threshold makes them secure in short circuit faults. As seen in Figures 12 and 13 the ROCOF and ROCPAD value in ROCOF-threshold2 and ROCPAD-threshold2 relays does not exceed the threshold. The trip signal of these relays is equal to zero. So, the output of these two relays with a logical OR gate is zero. Finally, the output signal of the proposed algorithm is equal to zero, and it does not work in short circuit faults.

The trip signal of relays and the proposed algorithm is shown in Figure 15. It can be seen that the trip signal of the proposed algorithm is zero. So, the proposed algorithm does not work in short circuit faults.

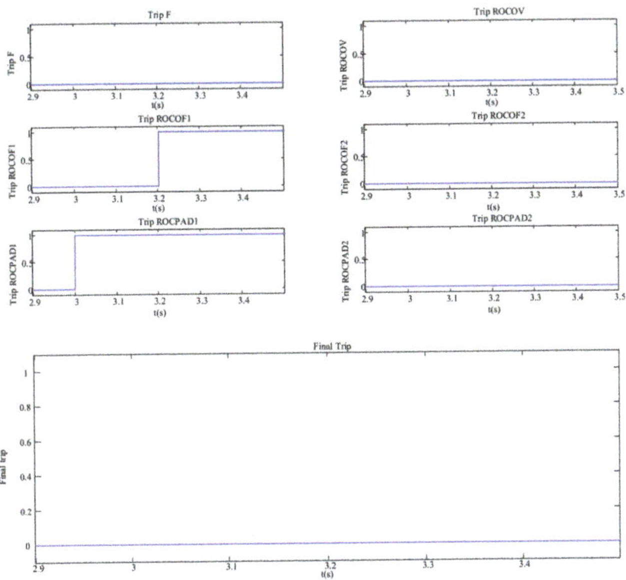

Figure 15. Trip signal of relays and the proposed algorithm in short circuit fault.

3.3. *Islanding with Zero Power Mismatch*

In the third case study, islanding mode happens with 0% power mismatch at t = 3. The trip signal of relays and the proposed algorithm is shown in Figure 16.

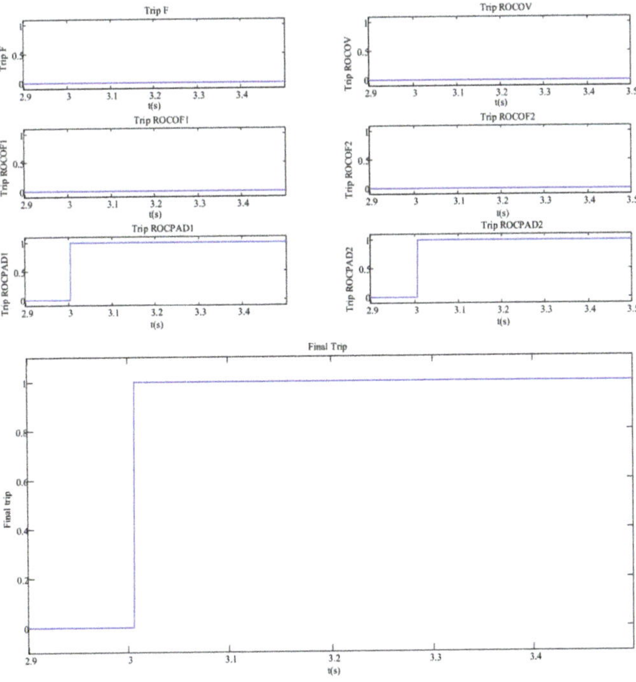

Figure 16. Trip signal of relays and the proposed algorithm in islanding mode with 0% power mismatch.

The frequency and voltage are not significantly low regarding power mismatch. The ROCOV, ROCOF, and OF/UF relays do not exceed their threshold. So, these relays do not detect islanding mode with low power mismatch. The ROCPAD relays can detect islanding mode even in 0% power mismatch. As seen in Figure 16, the trip signal of the ROCOV, ROCOF, and OF/UF is zero. The ROCPAD-threshold1 relay finds the islanding mode sooner than the ROCPAD-threshold2. The operation of ROCPAD-threshold1 is limited by ROCOV, ROCOF, and OF/UF relays. So, when the ROCPAD-thereshold2 exceeds its threshold, the proposed algorithm operates, and the generator will be separated from the power grid.

3.4. Load Shedding

In the fourth case study, at bus 7, a 1 MVA load with a power factor of 0.8 is disconnected from the power network at t = 3 s. The trip signal of relays and the proposed algorithm is shown in Figure 17.

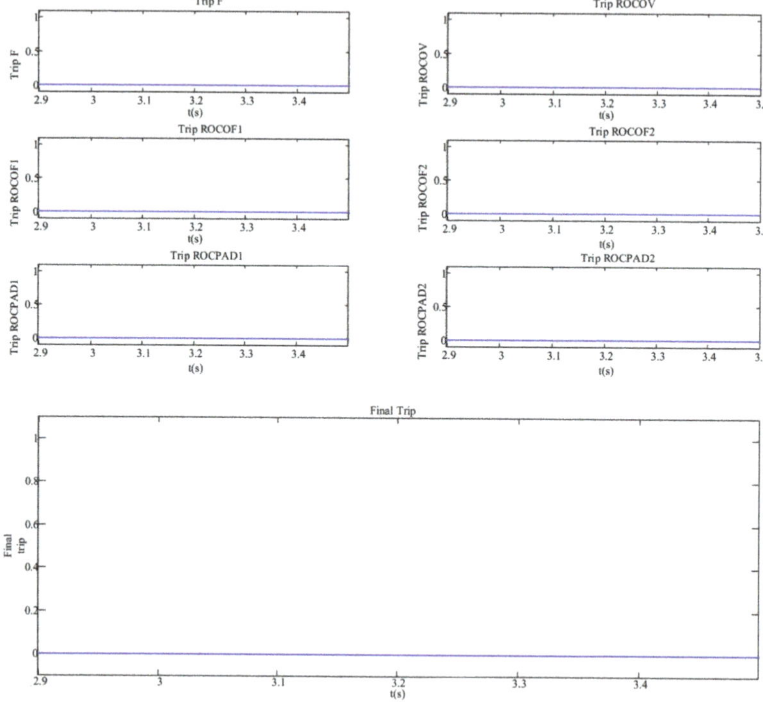

Figure 17. Trip signal of relays and the proposed algorithm in load shedding.

As seen in Figure 17, load shedding does not affect the operation of the relays. Thus, the output of the proposed algorithm is zero.

3.5. Capacitor Bank Switching

In the fifth case study, at bus 13, a 1.2 MVAr capacitor is connected to the power network at t = 3 s. The trip signal of relays and the proposed algorithm is shown in Figure 18.

As seen in Figure 18, capacitor switching does not affect the operation of the relays. Thus, the output of the proposed algorithm is zero.

3.6. Motor Switching

In the sixth case study, at bus 11, a 2.2 MW motor is connected to the power network at t = 3 s. The trip signal of relays and the proposed algorithm is shown in Figure 19. As seen in Figure 19, due to starting, the inrush current the voltage of PCC decreases. Therefore, the ROCOV value exceeds the pre-defined threshold. In this situation the ROCOV relay has mal-operation, but the proposed method operates correctly.

The proposed algorithm has been investigated in different islanding mode, different short circuit faults with different fault resistance, different load connecting, load shedding and capacitor bank disconnecting. The results show that the proposed algorithm separates clearly islanding mode from the non-islanding mode. Table 1 shows the comparison of detection time between the proposed algorithm and classic algorithms at different power mismatches. It can be seen that the proposed algorithm detects the island condition in less than 6 ms time delay. Moreover, in all cases, the speed of the proposed algorithm is faster than the classic algorithm.

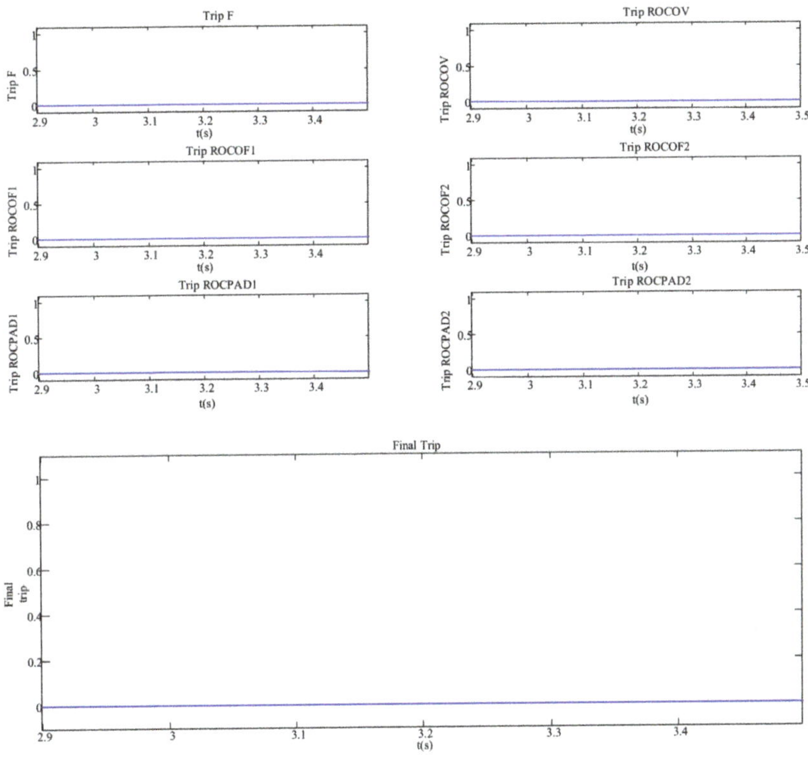

Figure 18. Trip signal of relays and the proposed algorithm in capacitor switching.

Table 1. Result of islanding detection using the proposed algorithm and classic algorithms for different active and reactive power mismatches.

Active Power Mismatch (Δ*P*)	Reactive Power Mismatch (Δ*Q*)	Detection Time (ms)			
		The Proposed Method	UF/OF	ROCOF	ROCPAD
80%	60%	3.21	3.7	3.21	3.56
60%	45%	3.92	4.1	3.92	4.2
40%	25%	4.13	4.5	4.13	4.2
20%	15%	5.22	7	6.5	5.22
15%	12%	5.37	10	7	5.37
10%	8%	5.56	NOT detect	12	5.56
5%	2%	5.89	NOT detect	NOT detect	5.89
3%	1.5%	5.91	NOT detect	NOT detect	5.91
2%	1%	5.94	NOT detect	NOT detect	5.94
0.5%	0%	5.98	NOT detect	NOT detect	5.98
0%	0%	6.00	NOT detect	NOT detect	6.00
−0.5%	−0.5%	5. 94	NOT detect	NOT detect	5.94
−3%	−1%	5.93	NOT detect	NOT detect	5.93
−5%	−2%	5.91	NOT detect	NOT detect	5.91
−10%	−12%	5.25	NOT detect	12.5	5.25
−20%	−15%	4.81	6.5	5.7	4.81
−40%	−25%	4.20	4.4	4.3	4.20
−60%	−45%	3.73	3.9	3.37	3.58
−80%	−60%	3.42	3.7	3.42	3.75
−85%	−70%	2.11	2.25	2.11	2.45

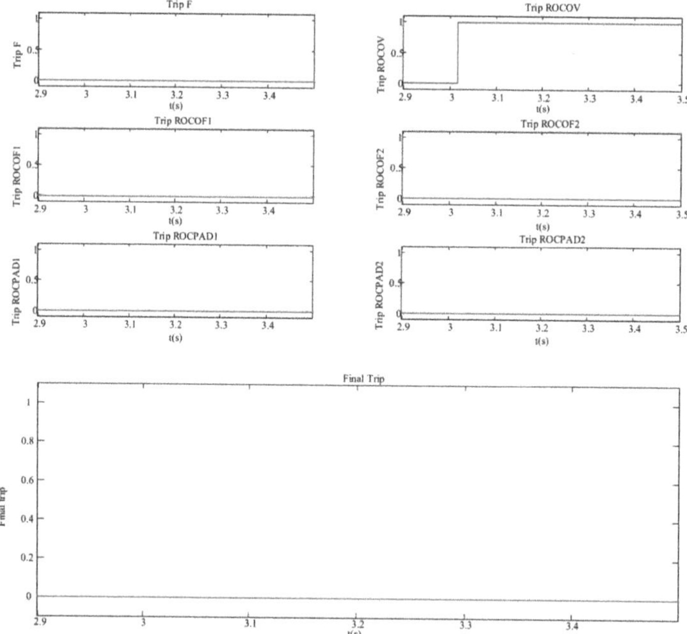

Figure 19. Trip signal of relays and the proposed algorithm in motor switching.

Table 2 shows the mal-operation of the proposed algorithm and classic algorithms in non-islanding events. It can be seen that the proposed algorithm, unlike classic algorithms, has not maloperation on non-islanding events.

Table 2. Result of operation of the proposed method and classic methods in the non-islanding event.

Grid-Connected Disturbances		Operation			
Event	Description	The Proposed Method	UF/OF	ROCOF	ROCPAD
Capacitor entry	1.2MVAR	non operation	non operation	non operation	non operation
	2.2MVAR	non operation	non operation	non operation	non operation
	3.3MVAR	non operation	non operation	non operation	non operation
	4.4MVAR	non operation	non operation	non operation	non operation
	5.5MVAR	non operation	non operation	non operation	mal-operation
Motor switching	2MW	non operation	non operation	non operation	non operation
	4MW	non operation	non operation	non operation	non operation
	6MW	non operation	non operation	non operation	non operation
	8MW	non operation	non operation	mal-operation	mal-operation
	10MW	non operation	non operation	mal-operation	mal-operation
faults	Single phase to ground	non operation	non operation	mal-operation	mal-operation
	Two phase	non operation	non operation	mal-operation	mal-operation
	Two phase to ground	non operation	non operation	mal-operation	mal-operation
	Three phase	non operation	non operation	mal-operation	mal-operation
Load shedding	1MVA	non operation	non operation	non operation	non operation
	2MVA	non operation	non operation	mal-operation	mal-operation
	3MVA	non operation	non operation	mal-operation	mal-operation
	4MVA	non operation	non operation	mal-operation	mal-operation

4. Conclusions

Islanding detection becomes very important by employing distributed generation in the power network. If the islanding is not detected correctly, it will damage to the network equipment. This paper presents a new algorithm for islanding detection in distributed generation. The proposed algorithm by combining different islanding detection methods presents a new method that detects islanding mode very fast. It does not suffer from drawbacks, which may be seen in other methods. It works effectively under a wide range of power mismatch. The proposed algorithm does not require much data. It detects islanding modes only by using the voltage and current at DG's locations. Mal-operation during short circuit faults is one of the major problems of most of the islanding detection methods. The proposed algorithm can distinguish between short circuit faults and islanding mode. The operation of the proposed algorithm was investigated in various conditions, including different short circuit faults with different fault resistance, different load connection and rejection and disconnection of compensator capacitor. In all cases, the proposed algorithm works correctly.

Author Contributions: All authors worked on this manuscript together. All authors read and approved the final manuscript.

Funding: This research received no external funding.

Conflicts of Interest: The authors declare no conflict of interest.

Appendix A

Table A1. The parameters of synchronous DG

Parameters	Value
Rate power	2 MVA
Voltage	380 V
Inertia constant (H)	1 s
X_d	1.44 pu

Table A2. The parameters of inverter-based DG

Power factor range	0.8 to 1 (lead and lag)
Nominal ac voltage	350 V rms
Maximum ac output current	1040 A rms
Nominal frequency	50/60 Hz
Reactive power range	±380 kVAr
Output power	473 KVA
Maximum PV operating current	1280 A
Suggested PV array peak power	725 Kw
Maximum array short circuit current at standard test conditions	1600 A
Maximum array short circuit current	2000 A
PV operating voltage range	510 to 850 V

References

1. Basso, T.; DeBlasio, R. IEEE 1547 Series of Standards: Interconnection Issues. *IEEE Trans. Power Electron.* **2004**, *19*, 1159–1162. [CrossRef]
2. Code, E.E. *Requirements for Grid Connection Applicable to All Generators*; ENTSO-E: Brussels, Belgium, 2013.
3. Tang, Y.; Li, F.; Zheng, C.; Wang, Q.; Wu, Y. PMU Measurement-Based Intelligent Strategy for Power System Controlled Islanding. *Energies* **2018**, *11*, 143. [CrossRef]
4. Almas, M.S.; Vanfretti, L. RT-HIL implementation of the hybrid synchrophasor and GOOSE-based passive islanding schemes. *IEEE Trans. Power Deliv.* **2015**, *31*, 1299–1309. [CrossRef]
5. Werho, T.; Vittal, V.; Kolluri, S.; Wong, S.M. A potential island formation identification scheme supported by PMU measurements. *IEEE Trans. Power Syst.* **2015**, *31*, 423–431. [CrossRef]
6. Xu, W.; Zhang, G.; Li, C.; Wang, W.; Wang, G.; Kliber, J. A Power Line Signaling Based Technique for Anti-Islanding Protection of Distributed Generators—Part I: Scheme and Analysis. *IEEE Trans. Power Deliv.* **2007**, *22*, 1758–1766. [CrossRef]
7. Bejmert, D.; Sidhu, T.S. Investigation Into Islanding Detection With Capacitor Insertion-Based Method. *IEEE Trans. Power Deliv.* **2014**, *29*, 2485–2492. [CrossRef]
8. Rostami, A.; Abdi, H.; Moradi, M.; Olamaei, J.; Naderi, E. Islanding detection based on ROCOV and ROCORP parameters in the presence of synchronous DG applying the capacitor connection strategy. *Electr. Power Compon. Syst.* **2017**, *45*, 315–330. [CrossRef]
9. Azim, R.; Li, F.; Xue, Y.; Starke, M.; Wang, H. An islanding detection methodology combining decision trees and Sandia frequency shift for inverter-based distributed generations. *IET Gener. Transm. Distrib.* **2017**, *11*, 4104–4113. [CrossRef]
10. Voglitsis, D.; Papanikolaou, N.; Kyritsis, A.C. Incorporation of harmonic injection in an interleaved flyback inverter for the implementation of an active anti-islanding technique. *IEEE Trans. Power Electron.* **2016**, *32*, 8526–8543. [CrossRef]
11. Wen, B.; Boroyevich, D.; Burgos, R.; Shen, Z.; Mattavelli, P. Impedance-based analysis of active frequency drift islanding detection for grid-tied inverter system. *IEEE Trans. Ind. Appl.* **2015**, *52*, 332–341. [CrossRef]
12. Emadi, A.; Afrakhte, H.; Sadeh, J. Fast active islanding detection method based on second harmonic drifting for inverter-based distributed generation. *IET Gener. Transm. Distrib.* **2016**, *10*, 3470–3480. [CrossRef]
13. Sun, Q.; Guerrero, J.M.; Jing, T.; Vasquez, J.C.; Yang, R. An islanding detection method by using frequency positive feedback based on FLL for single-phase microgrid. *IEEE Trans. Smart Grid* **2015**, *8*, 1821–1830. [CrossRef]
14. Liu, M.; Zhao, W.; Wang, Q.; Huang, S.; Shi, K. Compatibility Issues with Irregular Current Injection Islanding Detection Methods and a Solution. *Energies* **2019**, *12*, 1467. [CrossRef]
15. Du, P.; Nelson, J.; Ye, Z. Active anti-islanding schemes for synchronous-machine-based distributed generators. *IEE Proc. Gener. Transm. Distrib.* **2005**, *152*, 597–606. [CrossRef]
16. Roscoe, A.J.; Burt, G.M.; Bright, C.G. Avoiding the Non-Detection Zone of Passive Loss-of-Mains (Islanding) Relays for Synchronous Generation by Using Low Bandwidth Control Loops and Controlled Reactive Power Mismatches. *IEEE Trans. Smart Grid* **2014**, *5*, 602–611. [CrossRef]

17. Zamani, R.; Hamedani-Golshan, M.-E.; Alhelou, H.H.; Siano, P.; Pota, H.R. Islanding Detection of Synchronous Distributed Generator Based on the Active and Reactive Power Control Loops. *Energies* **2018**, *11*, 2819. [CrossRef]
18. Zamani, R.; Golshan, M.E.H.; Alhelou, H.H.; Hatziargyriou, N. A novel hybrid islanding detection method using dynamic characteristics of synchronous generator and signal processing technique. *Electr. Power Syst. Res.* **2019**, *175*, 105911. [CrossRef]
19. Tan, K.H.; Lan, C.W. DG System Using PFNN Controllers for Improving Islanding Detection and Power Control. *Energies* **2019**, *12*, 506. [CrossRef]
20. Vieira, J.C.; Freitas, W.; Xu, W.; Morelato, A. An Investigation on the Nondetection Zones of Synchronous Distributed Generation Anti-Islanding Protection. *IEEE Trans. Power Deliv.* **2008**, *23*, 593–600. [CrossRef]
21. Xu, W.; Freitas, W.; Huang, Z. A practical method for assessing the effectiveness of vector surge relays for distributed generation applications. *IEEE Trans. Power Deliv.* **2005**, *20*, 57–63.
22. Vieira, J.C.; Freitas, W.; Xu, W.; Morelato, A. Efficient Coordination of ROCOF and Frequency Relays for Distributed Generation Protection by Using the Application Region. *IEEE Trans. Power Deliv.* **2006**, *21*, 1878–1884. [CrossRef]
23. Samui, A.; Samantaray, S.R. Assessment of ROCPAD Relay for Islanding Detection in Distributed Generation. *IEEE Trans. Smart Grid* **2011**, *2*, 391–398. [CrossRef]
24. Zamani, R.; Golshan, M.E.H. Islanding Detection of Synchronous Machine-Based Distributed Generators Using Signal Trajectory Pattern Recognition. In Proceedings of the 6th International Istanbul Smart Grids and Cities Congress and Fair (ICSG), Istanbul, Turkey, 25–26 April 2018; pp. 91–95.
25. Pigazo, A.; Liserre, M.; Mastromauro, R.A.; Moreno, V.M.; Dell'Aquila, A. Wavelet-based islanding detection in grid-connected PV systems. *IEEE Trans. Ind. Electron.* **2008**, *56*, 4445–4455. [CrossRef]
26. Ning, J.; Wang, C. Feature Extraction for Islanding Detection Using Wavelet Transform-Based Multi-Resolution Analysis. In Proceedings of the 2012 IEEE Power and Energy Society General Meeting, San Diego, CA, USA, 22–26 July 2012; pp. 1–6.
27. Liu, X.; Zheng, X.; He, Y.; Zeng, G.; Zhou, Y. Passive Islanding Detection Method for Grid-Connected Inverters Based on Closed-Loop Frequency Control. *J. Electr. Eng. Technol.* **2019**, *1*, 1–10. [CrossRef]
28. Kong, X.; Xu, X.; Yan, Z.; Chen, S.; Yang, H.; Han, D. Deep learning hybrid method for islanding detection in distributed generation. *Appl. Energy* **2018**, *210*, 776–785. [CrossRef]
29. Abd-Elkader, A.G.; Saleh, S.M.; Eiteba, M.M. A passive islanding detection strategy for multi-distributed generations. *Int. J. Electr. Power Energy Syst.* **2018**, *99*, 146–155. [CrossRef]
30. Horowitz, S.H.; Phadke, A.G. *Power System Relaying*, 4th ed.; John Wiley & Sons: Hoboken, NJ, USA, 2013.

Article

Earth Fault Location Using Negative Sequence Currents

Amir Farughian *, Lauri Kumpulainen and Kimmo Kauhaniemi

School of Technology and Innovations, University of Vaasa, 6500 Vaasa, Finland; lauri.kumpulainen@uva.fi (L.K.); kimmo.kauhaniemi@uva.fi (K.K.)

* Correspondence: amir.farughian@uva.fi

Received: 27 August 2019; Accepted: 27 September 2019; Published: 30 September 2019

Abstract: In this paper, a new method for locating single-phase earth faults on non-effectively earthed medium voltage distribution networks is proposed. The method requires only current measurements and is based on the analysis of the negative sequence components of the currents measured at secondary substations along medium voltage (MV) distribution feeders. The theory behind the proposed method is discussed in depth. The proposed method is examined by simulations, which are carried out for different types of networks. The results validate the effectiveness of the method in locating single-phase earth faults. In addition, some aspects of practical implementation are discussed. A brief comparative analysis is conducted between the behaviors of negative and zero sequence currents along a faulty feeder. The results reveal a considerably higher stability level of the negative sequence current over that of the zero sequence current.

Keywords: sequence components; earth fault location; negative sequence current

---

1. Introduction

Feeder automation is one of the distinguishing features of smart grids. It aims at developing self-healing systems, able to locate faults and perform isolation and supply restoration automatically. Reliable fault location and indication is the key to this functionality. For short circuit faults, there are established methods for locating them, whereas, for locating earth faults, there is not a universally accepted, reliable and cost-effective method in the market for isolated neutral or compensated networks. However, a number of methods have been put forward to address this matter. A comprehensive review of the state-of-the-art methods for locating single-phase earth faults in medium voltage (MV) distribution networks is provided in Reference [1].

There are different types of fault location methods. Impedance-based methods [2–7] work based on calculating the apparent impedance seen when looking into the line from an end (measuring point) during the fault condition [8]. Since the line length between the measuring point and the fault location is proportional to the calculated impedance, by knowing the line impedance per unit length, the fault distance can be estimated. In Reference [2], the fault location and resistance are estimated using an iterative method. In Reference [3], a model developed for high-impedance faults consisting of two antiparallel diodes is proposed. A parameter estimation procedure is performed which uses the proposed model along with voltage and current signals to estimate the fault distance and other parameters. In Reference [4], the whole load of the feeder is modelled as one equivalent load tap located where the voltage drop due to the fault is at a maximum. Two assumptions are made i.e., once the load tap is assumed to be located before the fault point and once it is assumed to be after the fault. By solving the equations corresponding to those assumptions, the fault distance is estimated in unearthed networks. The method presented in Reference [5] scans the data obtained from power quality monitors and relays to estimate the fault distance in terms of the line impedance. The concept introduced in Reference [4] can be adapted for compensated distribution networks as presented in Reference [6]. In Reference [7], a method for fault distance calculation for compensated networks is put forward which requires a fewer number of settings compared to the method presented in Reference [4]. Despite all the efforts, these impedance-based methods are subject to errors as the fault impedance is unknown. Moreover, to locate single-phase to ground faults, the phase to ground voltages must be available [4], which is cost prohibitive. Generally, these methods are more suitable in transmission lines.

Travelling-wave based methods [9–13] work based on the transient voltages and currents (impulses) which are generated at the fault location. These waves propagate away from the fault point in both directions towards the ends. As these types of waves travel at the speed of light, by measuring the wave and their reflection arrival times at either terminal of the line, the fault distance can be determined [8]. A concept to create travelling waves is proposed in Reference [9], which is based on earthing the neutral via a controlled thyristor that provides a short period of high fault current. The method in Reference [10] works based on frequency spectrum components of waves generated as a result of fault occurrence. Reference [11] is based on the wave velocity of zero and aerial mode components and their arriving timestamps. Some results from field tests are reported in Reference [12] indicating the limitations of applying travelling wave fault location on distribution networks. In Reference [13], the frequency information of travelling voltage waves is integrated with their time domain information using continuous wavelet transform to locate faults. However, the main problem when using travelling-wave based methods is that due to the large number of branches and laterals in distribution networks, the reflections captured at the beginning of the feeder, may come from different sources and not necessarily from the fault location only. This makes the use of traveling wave-based methods problematic in distribution networks.

Signal injection-based methods are another type of fault location method [14–16]. The idea is to inject a pulsating current into the neutral of the system following a fault occurrence. The injected signal is detectable only in the faulted feeder and from the injection point up to the fault point. Although, this method has been field tested, the results show that it can be used for fault resistances only up to 300 Ω (without voltage measurement). In addition, the use of this method is limited to only compensated networks where the injection device can be coupled with the Petersen coil.

In References [17–20], earth fault location using zero sequence components is proposed. Despite the simplicity this approach offers, it suffers from a shortcoming i.e., false indication in networks where the length of the faulty feeder is long and the fault occurs close to the beginning of the feeder. In addition, this method in the manner proposed in Reference [20], when applied to compensated networks, requires connection and disconnection of the resistor in parallel with the Petersen coil in the right time, which could cause complexity.

Despite all these efforts, earth fault location in isolated and compensated neutral distribution networks remains a problem for which there does not seem to be a widely adopted solution and thus it is worth studying. Fault indication and location is a more challenging task in isolated and compensated neutral networks because their fault currents are so low that the simple threshold detection will often lead to poor results. One solution is to utilize neutral voltage measurement, in addition to current measurement. However, this neutral voltage measurement is not usually preferred at the MV side due to additional costs [18]. Therefore, there is a need for an earth fault indication method which is based on current measurement only. On the other hand, recent developments in smart grids have enabled communication between different parts of the network. By taking advantage of this feature, better solutions can be achieved as measurements from different parts of the network can be utilized to locate the fault.

In this paper, a new method for identifying the faulted segment (the segment between two secondary substations on which fault has occurred) in case of a single-phase to ground fault is presented. It is purely based on current measurements and no voltage measurement is required, which can be considered as an advantage. The proposed method employs the negative sequence current components to locate the fault. First, In Section 2, the negative sequence current along a faulted feeder is analysed in depth to gain insights into its behaviour. Then, based on that analysis, the fault location procedure is established in Section 3. In Section 4, the effectiveness of the method is evaluated by means of simulation using various types of medium voltage distribution networks. Lastly, some aspects of implementing the proposed method are discussed in Section 5.

2. Theoretical Background of the Proposed Method

In this section, to gain some insight into the proposed fault location method, the negative sequence currents on different locations on a feeder following a fault occurrence are discussed. The focus is on the difference between the negative sequence currents before and after the fault point.

To have a better understanding of the fault current, an MV distribution network with one healthy feeder and one with a single-phase to ground fault on it is shown in Figure 1. Earth capacitances and the flow of currents following a fault occurrence are shown in the figure. Fault current, especially in isolated neutral networks, is determined by network capacitances and mostly by the undamaged parts of the network.

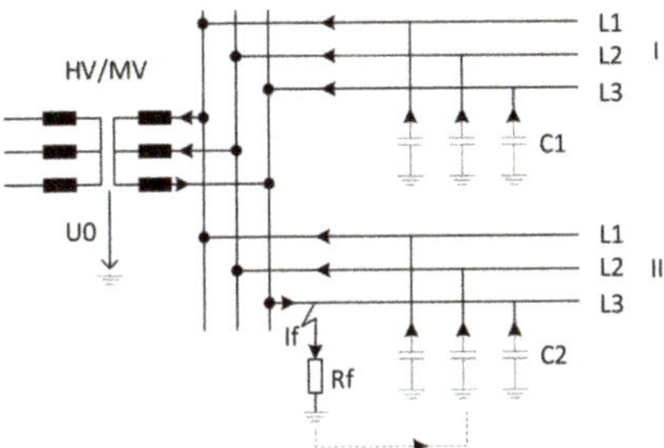

Figure 1. A single phase to ground fault in an isolated neutral system [21].

For studying the fault location problem with more details, the MV distribution network shown in Figure 2 is used. In this paper, the style of single line diagrams is derived from Reference [22]. A typical MV network consists of several feeders but from a theoretical viewpoint, only the faulty feeder needs to be studied with more details and others can be represented by a simple electrical equivalent circuit This aspect is considered in the figure using the single block "Background network" representing all the healthy feeders. A single-phase to ground fault occurs on the feeder under study between two secondary substations H and J at point F with the fault resistance R_F. The dashed lines indicate that the feeder can be of any length and can have any number of secondary substations. In Figure 2a, two arbitrary secondary substations are shown which indicates that the following analysis is valid for any point on the feeder. The goal here is to analyse the negative sequence current at point B, which is before the fault point and the negative sequence current after the fault location at point A. Note that points B and A are not electrically the same location. This is shown, for clarification, in Figure 2b.

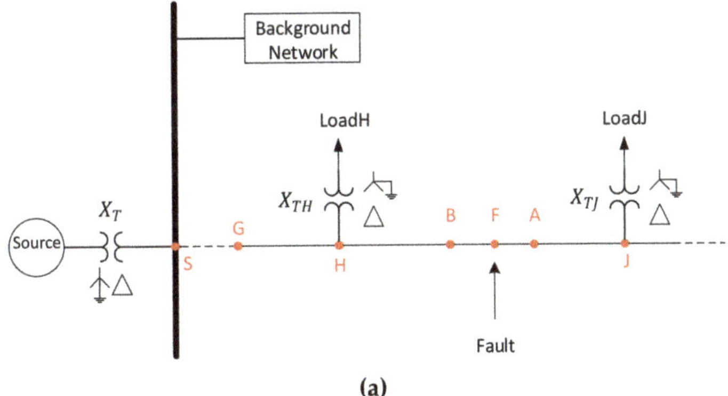

Figure 2. *Cont.*

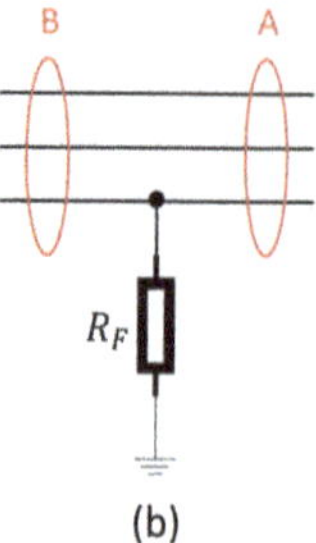

Figure 2. Single line diagram of a distribution network. (**a**) two arbitrary secondary substations are shown which indicates that the following analysis is valid for any point on the feeder. (**b**) B and A are not electrically the same location.

The sequence equivalent networks of the MV distribution system shown in Figure 2 can be obtained, as shown in Figures 3–5 (see also the notations list presented at the beginning of the paper). Note that line (to earth) capacitances only appear in the zero sequence network. In case of a single phase to ground fault, the sequence networks will be connected in series. The complete sequence network of the faulted network under study is shown in Figure 6. It can be simplified, as shown in Figure 7 where Z_G and Z_J are equivalent impedances of the parts marked on Figure 6. The currents $\overline{I_B}$ and $\overline{I_A}$ are the fundamental frequency components of the negative sequence currents at points B (before the fault) and A (after the fault) in Figure 2, respectively.

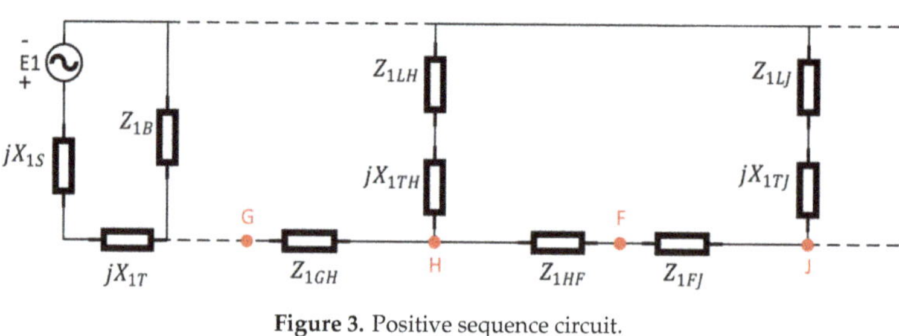

Figure 3. Positive sequence circuit.

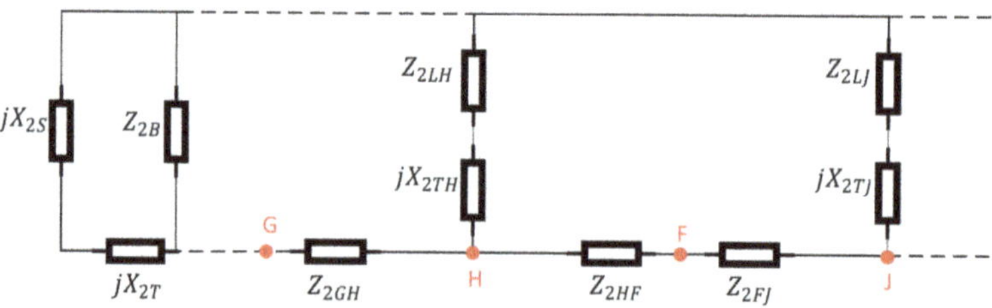

Figure 4. Negative sequence circuit.

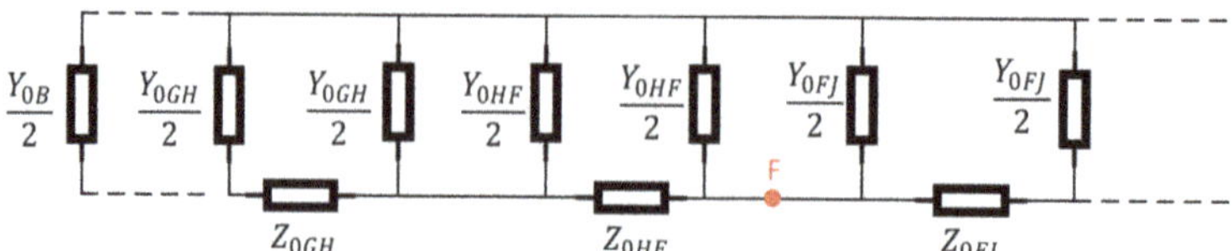

Figure 5. Zero sequence circuit.

The values of load impedances are normally rather large compared to the system impedances, so that they have a negligible effect on the faulted phase current. Therefore, it is common practice in fault analysis to neglect load impedances for shunt faults [22]. However, in the following, both scenarios are discussed, i.e., one where the effect of load is neglected and another where its effect is considered.

2.1. With No Load

Consider Figure 6. Shunt branches consist of transformers' reactance in series with load impedances. With no load assumption, the values of all the load impedances in the shunt branches are infinite and hence no current flows through the shunt branches in the positive and negative sequence networks. As a result, the negative sequence current measured at any point from beginning of the feeder to the fault point is constant and it is zero after the fault point. Therefore:

$$I_B > I_A \tag{1}$$

where, I_B and I_A are the magnitudes of the phasors $\overline{I_B}$ and $\overline{I_A}$, respectively.

2.2. With Load

In Figure 6, when taking load impedances into account, the negative sequence current measured along the feeder from S to F is not constant anymore as some currents will flow through the shunt branches as well. In addition, similarly, due to the flow of current through shunt branches after the fault point, the negative sequence current after the fault point is not zero anymore.

In practice, the series impedances in Figure 6 are negligible when compared to the shunt impedances, with the exception of Z_{2B} and X_{2s}. Indeed, Z_{2B} is comparatively low as it is the equivalent negative sequence impedance of several feeders i.e., the background network.

Now, consider Figure 7 which shows the simplified network where Z_G and Z_J are the total negative sequence impedances before and after the fault point, respectively. In Figure 7, the following is valid.

$$\bar{I}_B = \bar{I}_A + \bar{I}_F \tag{2}$$

$$\bar{I}_B = \frac{\overline{V}_f}{Z_G + Z_{2HF}} \tag{3}$$

$$\bar{I}_A = -\frac{\overline{V}_f}{Z_{2FJ} + Z_J} \tag{4}$$

As Z_G represents the part of the network which consists of negligible series impedances and the shunt branch Z_{2B}, it is a comparatively low impedance. In contrast, Z_J is a comparatively high impedance so that:

$$abs(Z_G + Z_{2HF}) < abs(Z_{2FJ} + Z_J) \tag{5}$$

Equations (3)–(5) yield:

$$I_B > I_A \tag{6}$$

Therefore, the negative sequence current before the fault point is higher than the negative sequence current after the fault point in both scenarios i.e., with and without considering the load.

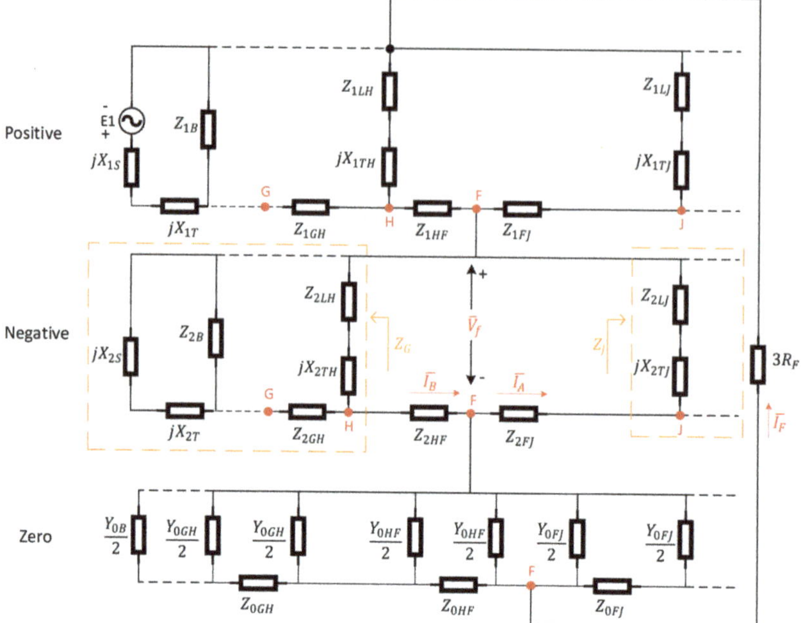

Figure 6. Sequence networks and interconnections for a phase-to-ground fault in the system of Figure 2.

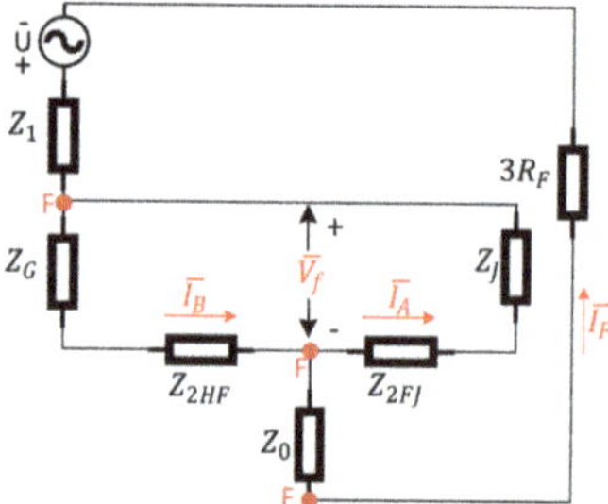

Figure 7. Simplified form of the network shown in Figure 6.

3. Outline of the Implementation of the Method

Considering the theory discussed in the previous section, when a single-phase-to-ground fault occurs, the negative sequence current exists and varies only a little in the section between the primary substation and the fault location whereas after the fault point, it is very low. This is simply illustrated in Figure 8.

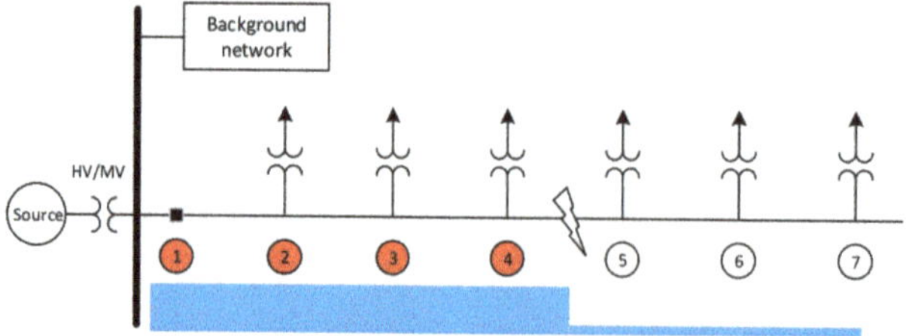

Figure 8. Negative sequence current magnitude along the feeder following a single-phase to ground fault occurrence.

Therefore, the proposed new fault location procedure is structured as follows:

1. The (change in the) negative sequence current at secondary substations (ideally at every substation) is obtained, following a fault occurrence.
2. These values are compared with a pre-set threshold at secondary substations.
3. Only from those secondary substations in which the threshold is exceeded, the fault detection information is sent to the control room.
4. The faulted segment is identified as the section between the last secondary substation that sends the fault detection information and the first secondary substation from which comes no signal.

For example, in Figure 8, the fault information is sent to the control room from those measuring points at which the magnitude of the negative sequence current has exceeded the threshold i.e., measuring points 1–4. The last three secondary substations (points 5–7) send no signal to the control room and therefore the faulted segment is identified as the one linking measuring points 4 and 5.

4. Simulation Results

To evaluate the validity of the proposed method, a set of simulations was carried out with PSCAD™/EMTDC™. The effectiveness of the proposed method was studied with different fault resistances in each of the following network types:

- Cabled urban compensated network;
- Cabled urban isolated neutral network;
- Rural compensated network;
- Rural isolated neutral network.

The network shown in Figure 9 is an urban compensated medium voltage distribution network with compensation degree 0.95. The voltage level in the medium voltage side is 20 kV which is fed from a 110-kV supply network using a 40 MVA transformer. The length of the feeder under study is 5.4 km consisting of AHXAMK underground cables. The feeder under study has five secondary substations equipped with current measurements. The current measurement points 2–6 are the secondary substations along the feeder, points 1 is at the beginning of the feeder and H refers to the measuring point at the beginning of an adjacent healthy feeder. A single-phase-to-earth fault, with the resistance of 30 Ω, occurs at 2.5 km from the beginning of the feeder.

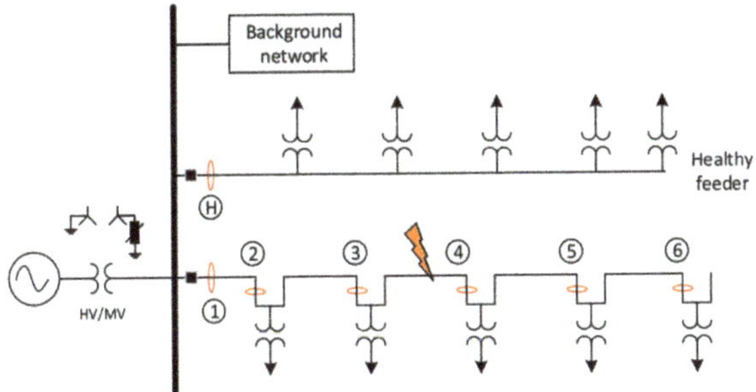

Figure 9. Cabled urban compensated network.

The magnitudes of the fundamental frequency phasors of the negative sequence currents with respect to time for faulty and healthy feeders are shown in Figure 10. The stabilized values obtained at $t = 0.2$ s are shown in a separate figure (Figure 11). Using the proposed method, one can readily determine the faulted section i.e., the one between secondary substations with measuring points 3 and 4. There is a considerable difference in the amplitude of the negative sequence currents before and after the fault points so that the negative sequence current is practically negligible after the fault point. It is worth noticing that the negative sequence current of the healthy feeder behaves in a similar way as that of the points after the fault point.

In addition, the variations of the negative sequence currents along the feeder from the beginning of it up to the fault point are so insignificant that the first three graphs (corresponding to locations before the fault point) are almost matched. This highlights the strength of using the negative sequence current in locating faults.

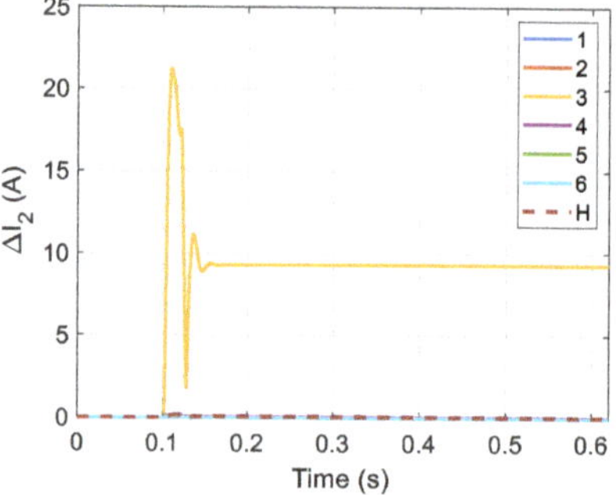

Figure 10. Negative sequence current magnitude.

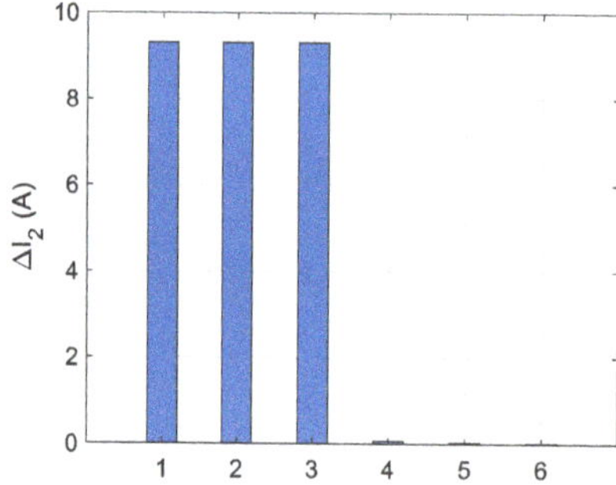

Figure 11. Negative sequence current magnitude.

To enable a comparison between negative and zero sequence currents, similar types of graphs are obtained for the zero sequence currents and shown in Figures 12 and 13. Contrary to the negative sequence current, the zero-sequence current varies considerably along the feeder and is not negligible after the fault point. This reveals the shortcoming of using the zero-sequence current over the negative sequence current.

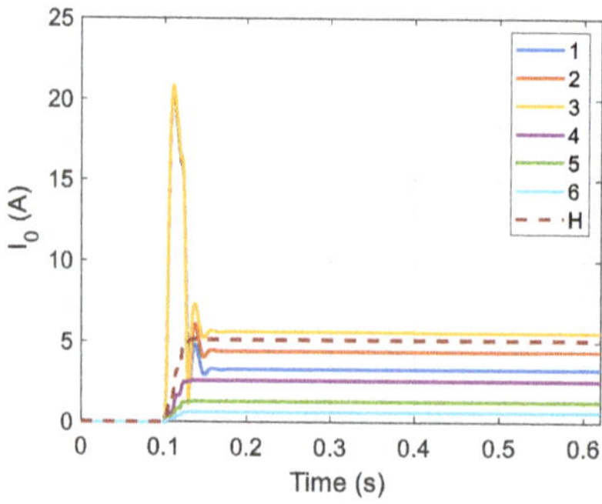

Figure 12. Zero sequence current magnitude.

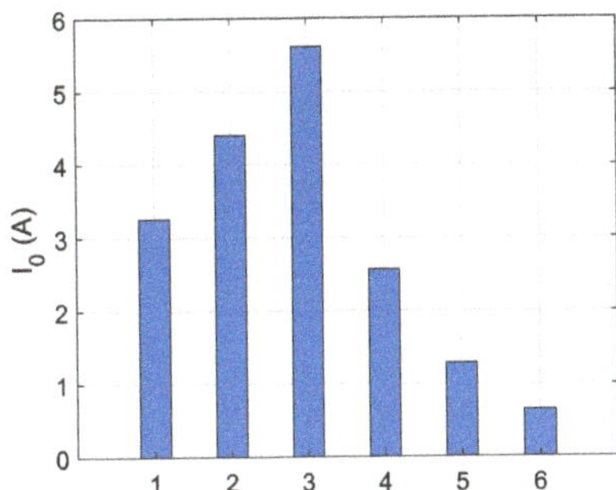

Figure 13. Zero sequence current magnitude.

To further investigate the validity and performance of the proposed method, five fault resistances are studied ranging from 0.01 Ω up to 1 kΩ. The changes in the negative sequence currents, caused by the fault, at the beginning of the healthy and faulty feeders and at each secondary substation of the faulty feeder are provided in Table 1. The choice of the threshold might vary from network to network. Developing a procedure to decide the proper threshold for a given network is not the focus of this paper. For simulations presented in this paper, the threshold is chosen to be 3 A for urban networks and 2 A for rural networks. For fault resistances between 0.01 Ω and 1000 Ω, it is clear from the table that for points 1 to 3, negative sequence currents exceed the threshold (3 A) whereas these values are below the threshold for points 4 to 6. Therefore, the faulted segment is the one between the second and third secondary substations i.e., between points 3 and 4.

Table 1. Negative sequence currents in urban compensated network.

Rf (Ω)	ΔIn1 (A)	ΔIn2 (A)	ΔIn3 (A)	ΔIn4 (A)	ΔIn5 (A)	ΔIn6 (A)	ΔInH (A)
0.01	9.57	9.57	9.57	2.6	0.04	0.02	0.08
1	9.48	9.48	9.49	2.6	0.04	0.02	0.08
10	9.41	9.41	9.41	2.59	0.04	0.02	0.08
100	8.81	8.81	8.82	2.42	0.03	0.02	0.07
1000	3.38	3.37	3.38	0.92	0.01	0.01	0.03

The same procedure can be applied to isolated networks, as shown in Figure 14. The changes in the negative sequence currents, following the fault occurrence, are provided in Table 2. Using the table and the same 3 A threshold, the faulted segment can be determined successfully for all the 5 fault resistances.

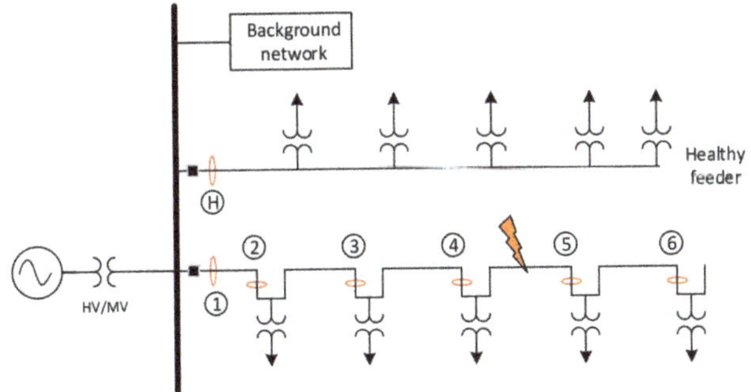

Figure 14. Urban isolated network.

Table 2. Negative sequence currents in urban isolated network.

Rf (Ω)	ΔIn1 (A)	ΔIn2 (A)	ΔIn3 (A)	ΔIn4 (A)	ΔIn5 (A)	ΔIn6 (A)	ΔInH (A)
0.01	58.91	58.9	58.94	56.54	0.26	0.13	0.48
1	58.89	58.88	58.92	56.51	0.26	0.13	0.48
10	58.06	58.06	58.09	55.72	0.25	0.13	0.48
100	31.73	31.73	31.75	30.45	0.14	0.07	0.26
1000	3.85	3.85	3.85	3.68	0.02	0.01	0.03

Now consider the rural compensated network shown in Figure 15. The first 5 km of the feeder under study consists of cables and the rest 45 km is overhead lines. The cable and overhead lines types modelled in the simulations are shown on the figure. A single-phase to ground occurs at 28 km from the beginning of the feeder. Similarly, using a threshold of 2 A and Table 3, the faulted segment is successfully identified, as the one between the second and third secondary substations (points 3 and 4).

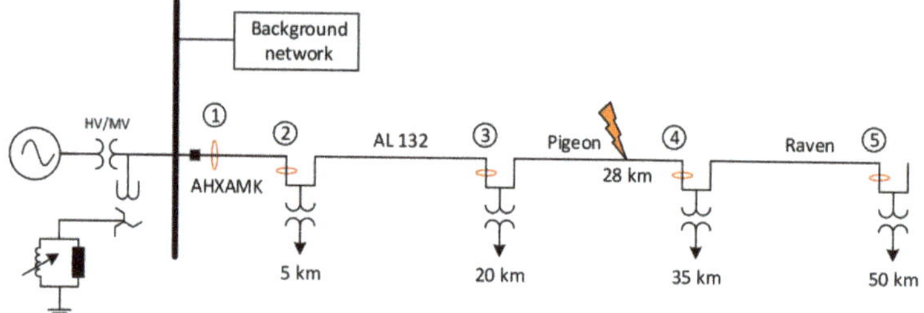

Figure 15. Rural compensated network.

Table 3. Negative sequence currents in rural compensated network.

Rf (Ω)	ΔIn1 (A)	ΔIn2 (A)	ΔIn3 (A)	ΔIn4 (A)	ΔIn5 (A)
0.01	5.25	5.25	5.26	0.59	0.29
1	5.25	5.24	5.26	0.59	0.29
10	5.21	5.21	5.22	0.59	0.29
100	4.94	4.94	4.95	0.56	0.27
1000	2.39	2.38	2.39	0.27	0.13

Lastly, in Figure 16, a rural isolated neutral network is shown. The change in the magnitude of the negative sequence currents for the points marked on the figure are shown in Table 4. In a similar fashion, the faulted segment is determined as the one between points 3 and 4 using the same 2 A threshold.

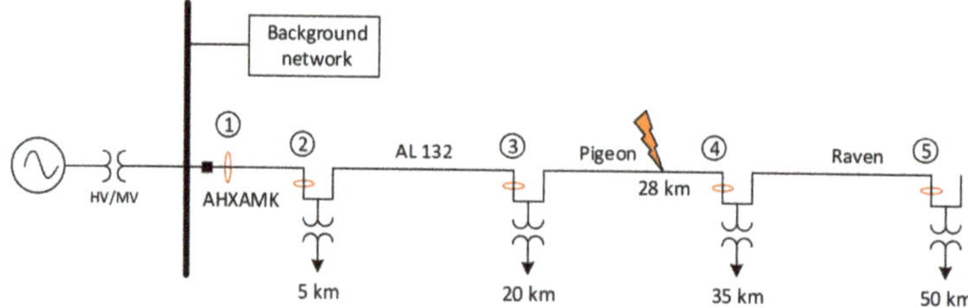

Figure 16. Rural isolated network.

Table 4. Negative sequence currents in rural isolated network.

Rf (Ω)	ΔIn1 (A)	ΔIn2 (A)	ΔIn3 (A)	ΔIn4 (A)	ΔIn5 (A)
0.01	15.21	15.19	15.23	1.72	0.85
1	15.21	15.19	15.23	1.72	0.85
10	15.16	15.13	15.18	1.72	0.84
100	13.4	13.37	13.41	1.52	0.74
1000	3	3	3.01	0.34	0.17

5. Discussion

In practice, current measurements at each secondary substation can be made using Rogowski coils, which are also suitable for retrofit installations. In the tables presented in Section 4, there are current values which are very low. It should be noted that the accuracy of the negative sequence current depends on the accuracy of CTs or current sensors (Rogowski coils) used and that decides the lowest possible value of the negative sequence current that can be reliably measured.

When a fault occurs on a distribution feeder, the information on the fault conditions must be delivered to the control room where SCADA system or DMS processes the information to visualize the fault location to the operator or to generate an automatic FLIR (fault location, isolation and restoration) switching sequence. The proposed method requires some level of communication between secondary substations and the primary substation. However, as communication will be widespread in future distribution networks, the proposed method is worth considering.

In an ideal symmetrical network, the negative zero sequence measured at any point is zero under normal condition. However, in practice, there is some level of negative sequence current even under no fault condition, for example due to asymmetry of loads. Therefore, it is important to note that the proposed method is based on the change in the negative sequence current and not the negative sequence current alone.

Compared to the current injection method presented in Reference [14], the proposed method in this paper has a clear advantage as there is no need for any extra injecting device. In addition, the pulse injection method works only with compensated neutral networks where the injection device can be connected in parallel with the Petersen coil. For isolated neutral networks where there is no possibility for connecting the injection device, the pulse injection method cannot be used.

In case of a single-phase-to-ground fault, unlike the zero sequence current, the negative sequence current remains almost constant with negligible variations along the feeder as shown in Section 4. Moreover, in compensated networks, the difference in the zero sequence current level for points before and after the fault could be too low for finding suitable and reliable criteria for locating the fault. Therefore, there is a need to increase this deference level somehow. Reference [20] suggests to connect and control an additional resistor in parallel with the Petersen coil. However, that action may lead to more complexity when it comes to practical implementation. In addition, there is a risk that in case of a long feeder, the line to ground capacitances of that part of the feeder, which is located after the fault point, provide a considerable amount of zero sequence current so that the zero-sequence based method fails to operate. This highlights the advantage of using the negative sequence current over the zero sequence current in locating faults in distribution networks.

In summary, the main advantages the negative sequence current based method offers are:

- No voltage measurement is required;
- Applicable to both isolated and compensated distribution networks;
- No need for any additional injection device or resistance.

6. Conclusions

Earth fault location on MV distribution networks using the negative sequence currents in the manner proposed in this paper appears to be promising. The proposed method is applicable to both compensated and isolated neutral networks. When using this method, only current measurements are required and not voltage measurements, which is an advantage. The theoretical reasoning to justify the proposed method was discussed in detail and simulation results obtained from different types of practical MV distribution networks confirmed the validity and performance of the method. Some further studies are still needed to address the aspects relating to practical implementation and for example the effects of decentralized compensation.

Author Contributions: Conceptualization, L.K.; Formal analysis, A.F.; Funding acquisition, K.K.; Investigation, A.F.; Project administration, K.K.; Software, A.F.; Supervision, K.K.; Writing—original draft, A.F.; Writing—review & editing, L.K. and K.K.

Funding: This work was supported by the European Regional Development Fund (ERDF) (Business Finland grant No. 4332/31/2014 and Council of Tampere Region grant No. A73094).

Conflicts of Interest: The authors declare no conflict of interest.

List of Notations

X_{0S}	Zero sequence source reactance
X_{1S}	Positive sequence source reactance
X_{2S}	Negative sequence source reactance
X_{0T}	Zero sequence reactance of the main transformer
X_{1T}	Positive sequence reactance of the main transformer
X_{2T}	Negative sequence reactance of the main transformer
Z_{0GH}	Zero sequence impedance of the feeder between nodes G and H
Z_{1GH}	Positive sequence impedance of the feeder between nodes G and H
Z_{2GH}	Negative sequence impedance of the feeder between nodes G and H
X_{0TH}	Zero sequence reactance of transformer H
X_{1TH}	Positive sequence reactance of transformer H
X_{2TH}	Negative sequence reactance of transformer H
Z_{0LH}	Zero sequence impedance of load H
Z_{1LH}	Positive sequence impedance of load H
Z_{2LH}	Negative sequence impedance of load H
Z_{0HF}	Zero sequence impedance of the feeder between node H and the fault location
Z_{1HF}	Positive sequence impedance of the feeder between node H and the fault location
Z_{2HF}	Negative sequence impedance of the feeder between node H and the fault location
R_F	Fault resistance
Z_{0FJ}	Zero sequence impedance of the feeder between the fault location and node J
Z_{1FJ}	Positive sequence impedance of the feeder between the fault location and node J
Z_{2FJ}	Negative sequence impedance of the feeder between the fault location and node J
X_{0TJ}	Zero sequence reactance of transformer J
X_{1TJ}	Positive sequence reactance of transformer J
X_{2TJ}	Negative sequence reactance of transformer J
Z_{0LJ}	Zero sequence impedance of load J
Z_{1LJ}	Positive sequence impedance of load J
Z_{2LJ}	Negative sequence impedance of load J
Y_{0B}	Equivalent zero sequence admittance representing the phase to earth capacitances of the background network
Z_{1B}	Equivalent positive sequence impedance of the background network
Z_{2B}	Equivalent negative sequence impedance of the background network
Y_{0GH}	Zero sequence admittance representing phase to earth capacitances between nodes G and H
Y_{0HF}	Zero sequence admittance representing phase to earth capacitances between nodes H and the fault location
Y_{0FJ}	Zero sequence admittance representing phase to earth capacitances between the fault location and node J

References

1. Farughian, A.; Kumpulainen, L.; Kauhaniemi, K. Review of methodologies for earth fault indication and location in compensated and unearthed MV distribution networks. *Electr. Power Syst. Res.* **2018**, *154*, 373–380. [CrossRef]
2. Novosel, D.; Hart, D.; Hu, Y.; Myllymaki, J. System for locating faults and estimating fault resistance in distribution networks with tapped loads. U.S. Patent No. 5,839,093, 17 November 1998.
3. Iurinic, L.U.; Herrera-Orozco, A.R.; Ferraz, R.G.; Bretas, A.S. Distribution Systems High-Impedance Fault Location: A Parameter Estimation Approach. *IEEE Trans. Power Deliv.* **2016**, *31*, 1806–1814. [CrossRef]
4. Altonen, J.; Wahlroos, A. Advancements in fundamental frequency impedance based earth fault location in unearthed distribution systems. In Proceedings of the CIRED 19th International Conference on Electricity Distribution, Vienna, Austria, 21–24 May 2007.
5. Kulkarni, S.; Santoso, S.; Short, T.A. Incipient Fault Location Algorithm for Underground Cables. *IEEE Trans. Smart Grid* **2014**, *5*, 1165–1174. [CrossRef]
6. Altonen, J.; Wahlroos, A.; Pirskanen, M. Advancement in earth-fault location in compensated MV-networks. In Proceedings of the CIRED 21st International Conference on Electricity Distribution, Frankfurt, Germany, 6–9 June 2011; p. 4.
7. Altonen, J.; Wahlroos, A. Novel algorithm for earth-fault location in compensated MV-networks. In Proceedings of the 22nd International Conference and Exhibition on Electricity Distribution (CIRED 2013), Stockholm, Sweden, 10–13 June 2013; pp. 1–4.
8. *IEEE Guide for Determining Fault Location on AC Transmission and Distribution Lines*; EEE Std C37.114-2014 (Revision of IEEE Std C37.114-2004); IEEE: New York, NY, USA, 2015; pp. 1–76.
9. Elkalashy, N.I.; Sabiha, N.A.; Lehtonen, M. Earth Fault Distance Estimation Using Active Traveling Waves in Energized-Compensated MV Networks. *IEEE Trans. Power Deliv.* **2015**, *30*, 836–843. [CrossRef]
10. Sadeh, J.; Bakhshizadeh, E.; Kazemzadeh, R. A new fault location algorithm for radial distribution systems using modal analysis. *Int. J. Electr. Power Energy Syst.* **2013**, *45*, 271–278. [CrossRef]
11. Liang, R.; Fu, G.; Zhu, X.; Xue, X. Fault location based on single terminal travelling wave analysis in radial distribution network. *Int. J. Electr. Power Energy Syst.* **2015**, *66*, 160–165. [CrossRef]
12. Thomas, D.W.P.; Carvalho, R.J.O.; Pereira, E.T.; Christopoulos, C. Field trial of fault location on a distribution system using high frequency transients. In Proceedings of the 2005 IEEE Russia Power Tech, St. Petersburg, Russia, 27–30 June 2005; pp. 1–7.
13. Borghetti, A.; Bosetti, M.; Nucci, C.A.; Paolone, M.; Abur, A. Integrated Use of Time-Frequency Wavelet Decompositions for Fault Location in Distribution Networks: Theory and Experimental Validation. *IEEE Trans. Power Deliv.* **2010**, *25*, 3139–3146. [CrossRef]
14. Druml, G.; Raunig, C.; Schegner, P.; Fickert, L. Fast selective earth fault localization using the new fast pulse detection method. In Proceedings of the 22nd International Conference and Exhibition on Electricity Distribution (CIRED 2013), Stockholm, Sweden, 10–13 June 2013; pp. 1–5.
15. Tengg, C.; Schmaranz, R.; Schoass, K.; Marketz, M.; Druml, G.; Fickert, L. Evaluation of new earth fault localization methods by earth fault experiments. In Proceedings of the 22nd International Conference and Exhibition on Electricity Distribution (CIRED 2013), Stockholm, Sweden, 10–13 June 2013; pp. 1–4.
16. Raunig, C.; Fickert, L.; Obkircher, C.; Achleitner, G. Mobile earth fault localization by tracing current injection. In Proceedings of the 2010 Electric Power Quality and Supply Reliability Conference, Kuressaare, Estonia, 16–18 June 2010; pp. 243–246.
17. Kumpulainen, L.; Pettisalo, S.; Sauna-aho, S. *A Secondary Substation Monitoring Based Method for Earth-Fault Indication in MV Cable Networks*; Aalto University: Espoo, Finland, 2011.
18. Lehtonen, M.; Siirto, O.; Abdel-Fattah, M.F. Simple fault path indication techniques for earth faults. In Proceedings of the 2014 Electric Power Quality and Supply Reliability Conference (PQ), Rakvere, Estonia, 11–13 June 2014; pp. 371–378.
19. Horákl, M.; Vinklárek, P.; Kordas, J.; Grossmann, J. Earth fault location in compensated MV network using a handheld measuring device. *J. Eng.* **2018**, *2018*, 1281–1285. [CrossRef]
20. Wang, P.; Chen, B.; Zhou, H.; Cuihua, T.; Sun, B. Fault Location in Resonant Grounded Network by Adaptive Control of Neutral-to-Earth Complex Impedance. *IEEE Trans. Power Deliv.* **2018**, *33*, 689–698. [CrossRef]

21. Lakervi, E.; Partanen, J. *Sähkönjakelutekniikka*; Otatieto: Helsinki, Finland, 2004; ISBN 978-951-672-359-7.
22. Blackburn, L.J. *Symmetrical Components for Power Systems Engineering*; Electrical and Computer Engineering; CRC Press: Boca Raton, FL, USA, 1993; ISBN 978-0-8247-8767-7.

Article

Design of the Smart Grid Architecture According to Fractal Principles and the Basics of Corresponding Market Structure

Albana ILO

TU Wien—Institute of Energy Systems and Electrical Drives, 1040 Vienna, Austria; albana.ilo@tuwien.ac.at; Tel.: +43-1-58801-370114

Received: 7 October 2019; Accepted: 28 October 2019; Published: 31 October 2019

Abstract: Nowadays, there is a dramatic upsurge in the use of renewable energy resources, ICT and digitalization that requires more than the straightforward refinements of an established power system structure. New solutions are required to perform dynamic optimizations in real time, closed loops and so on, taking into account the high requirements on data privacy and cyber security. The *LINK*-paradigm was designed to meet these requirements. It was developed on the basis of the bottom-up method that can lead to misinterpretations or wrong conclusions. This work mainly deals with the verification of the authenticity and correctness of *LINK*. Fractal analysis is used to identify the unique and independent elements of smart grids required for the design of an architectural paradigm. The signature of the fractal structure, the so-called fractal pattern, is founded and referred to as electrical appliances (ElA). The latter has proven to be the key component of the architectural *LINK* paradigm. The definition of the *LINK* paradigm is finally validated: It consists of unique and independent elements that avoid misinterpretation or the need for any changes in its definition. Additionally, the fractal analysis indicates two fractal anomalies in the existing power system structure, while the fractal dimension calculation insinuates the highest complexity in the fractal level of electrical devices. The *LINK*-based holistic architecture is given a finishing touch. A compact presentation of the control chain strategy is provided that should facilitate its practical implementation. The basis for the harmonization of the market structure with the grid link arrangements is established. The processes of demand response and conservation voltage reduction are presented under the new findings.

Keywords: holistic power system architecture; smart grid; fractal design; fractal grid; *LINK*-paradigm; market design; local electricity market

---

1. Introduction

In the last 15 years many papers and books developed to investigate and design smart grids have been written. For the most part, concepts are introduced and various models are developed that lead to extremely ramified and complex schemas. Various projects are focused on certain parts of power systems without considering the integrity of smart grids. As a result, all efforts remain to the level of prototypes or isolated model regions.

So, what is the problem? "An extremely diverse and complex topic" is the answer. "... Each time we get into this logjam of too much trouble, too many problems, it is because the methods that we are using are just like the ones we have used before ... " teaches us Richard Feynman, the greatest [1]. This means the problem may be in is in the origin, in the definitions of the circulating smart grid concepts. Without elaborate concepts or paradigms characterized by unique and independent elements, serious design flaws and unclear operating procedures may result [2].

1.1. Literature Review on the Popular Concepts of Smart Grids

The traditional structure of power systems became questionable at the beginning of the liberalization; the market rules overcome the technical. Several blackouts and electricity crises were the consequence [3]. After the blackout that plagued the United States and Canada, on August 14th 2003, the term "smart grid" was introduced. It was mostly related to the increase of the transmission capacity and level of automation in the grid [4]. The electricity crisis in California (2000–2001) gave the rise of distributed generation (DG) a major boost [5]. The installation of the small photo voltaic (PV) plants on house roofs transformed consumers during the day into electricity producers. A new category of customers appeared: The prosumers. The meaning of the term smart grids have evolved and now stands for the modernization of power systems and meeting all the requirements of the time. According to [6], the scope of smart grids covers the entire power system (high, medium and low voltage level, HV, MV and LV, respectively) right down to the individual electrical appliances in customer plants (CP).

The need to design smart grids has brought onstage various smart grid concepts: Virtual power plants (VPP), microgrids, cellular approach (CA), web of cells (WoC).

The VPP concept as an aggregation of a number of distributed generators (DG) was first introduced in 2001 [7,8] to enable their participation in the electricity market. It soon turned out that an adaptation was necessary to take into account the voltage and frequency performance [9]. VPP optimization focuses on internal energy dispatch and external market participation [10]. The definition of VPP, in particular its technical aspect, continues to be in discussion [11–14]. The basic cell controller architecture [15] uses the VPP concepts. It has a layered control hierarchy that uses distributed agent technology. It defines three control modules: Local, regional and enterprise, requiring a tremendous amount of data to be exchanged between its modules [16]. Market researches focus on details of how they function, such as the development of different participation models in the balancing market [17] or the provision of secure market products [18], but not on the harmonization of the market structure with the power grid architecture.

Microgrids were developed with a focus on the technical issues of DG integration [19,20]. They lead to very complex architectures, setting the microgrids, nanogrids, etc., according to the Matryoshka-doll principle [21]. The detailed definition of microgrids is still under discussion in technical forums [22]. It focuses on distribution level and includes non-unique and undefined elements. "Load" is an important component of the microgrid definition that according to [23] has an ambiguous meaning. It can present the consumption of a device, a part of the grid or the total active and/or reactive power consumed by all devices of the power system. The scope of "the host power system" is undefined. All this complicates the definition of the size of microgrids and consequently the design of clear and standardized structures. The lack of a definitive definition gives to any project or initiative the opportunity to adapt the basic definition to their specific needs and develop individual processes. The latter cannot guarantee financial revenue streams, cannot be reliably audited, impedes pooling of multiple microgrid projects into a financial asset class and does not allow for wide-spread and attractive microgrid and distributed energy resource project deployment, [24].

CA can be seen as an attempt to further develop the microgrid concept taking into account the distribution and transmission grid. It is a self-controlled small microgrid, which is integrated with a modular smart grid ICT infrastructure [25]. The CA-architecture is based on five different energy cell types: Residential, commercial and industrial-plant, -area and –park. The distribution grid is considered through "connection corridors" while the transmission grid through "transmission corridors" [26,27].

WoC is the latest concept introduced into the landscape of the smart grid concepts. The cell definition has evolved very dynamically in a short time [28–32], is generic and does not appear unique in the literature [31,32]. In the cell-based decentralized control framework WoC, cells are defined as "non-overlapping topological subsets of a power system associated with a scale-independent operational responsibility" [31] without specifying the criteria for the practical definition of its size.

The design of market structure is handled separately, not in combination with the technical architecture to enable their harmonization [32–34]. There are also attempts to design the architecture of smart grids without using a particular concept [35,36].

As a result all architectures based on the VPP, microgrids or their combinations with VPP, CA and WoC are very complex and hardly practicable. The harmonization of the technical smart grid architectures with the market structures does not seem to be in focus yet.

An ultra-large scale comprehensive control framework is needed to avoid the growing and unmanageable patchwork of grid implementations that are not sustainable on a large scale [37]. *LINK*-paradigm enables the holistic consideration of the power system (HV, MV and LV levels) and CP [38] and of the market [39]. However, it was developed on the basis of the bottom-up method that can lead to misinterpretations or wrong conclusions. The goal of this paper is the verification of the authenticity and correctness of *LINK*. The identification of unique and independent construction elements of smart grids is done using fractality principles.

1.2. Literature Review on the Use of Fractal Analysis in Smart Grids

In the past few years, fractality is increasingly being used to describe complex structures in nature [40]. It has recently been considered as a useful tool for analyzing, understanding and designing smart grids. According to [41] the scale-invariant properties of power grids are due to their design. Self-similarity is already identified in load flow characteristics [42,43], power demand behaviour [44], fault diagnostics [45,46], and protective relays [47,48]. Similarities have also been discovered in blackouts of interconnected power grids and their sub grids [49]. Fractal principles have been studied in grid structures and fractal dimensions have been calculated for different grids [50,51] to develop a cluster power system philosophy with cluster network structure [52,53]. Other authors have organized a smart grid based on a fractal model and have associated it with the holonic concept and holarchy [54]. Others consider fractals as an instrument that facilitates the realization of smart grids in a decentralized fashion [55]. However, the identification of a fractal pattern describing smart grids in its entirety has not yet been a research focus.

Section 2 gives an overview of the methodology used in this work. The in-depth investigation of the fractality features throughout the smart grid structure, including the calculation of fractal dimension, is presented in Section 3.1. The identified signature of the fractal structure, the so-called fractal pattern, in smart grids constitutes the foundation of the *LINK*-paradigm and of the holistic model, described in Section 3.2. Section 3.3 presents the holistic architecture and some applications of the *LINK*-solution, including a compact presentation of the control chain strategy and some applications as the harmonization of the technical architecture with the market structure, demand response, etc. Finally, some concluding remarks are summarized.

2. Methodology

Figure 1 shows the methodology used to design the smart grid architecture. In the first phase a superordinate connection between some well-known operation processes as "volt/var control in medium voltage grids" was established and the holistic technical model was predicted [38]. According to the prediction, the holistic technical model should include the entire power system, i.e., high, medium and low voltage levels and the customer plants. The bottom-up or induction method was initially used to derive the paradigm of smart grids "*LINK*" and the corresponding holistic architecture. The latter have shown the fractality feature of similar structural details.

The definition of the architectural paradigm of smart grids should be clear and contain unambiguous elements so that experts or users cannot make different interpretations [2]. Since misinterpretations or wrong decisions are preprogrammed in the bottom-up method, the top-down or the deduction method is used to verify the authenticity of the paradigm *LINK* and to exclude any suspicion or misinterpretation. After a detailed analysis, the fractal pattern of smart grids is revealed. The architectural *LINK*-paradigm is confirmed: Among others, it consists of the fractal pattern. Thereafter, the holistic technical and market model and the *LINK*-based holistic architecture have been given a finishing touch. The new architecture enables the description and realization of all processes required for the operation of smart grids, such as demand response, load/generation balance and so on.

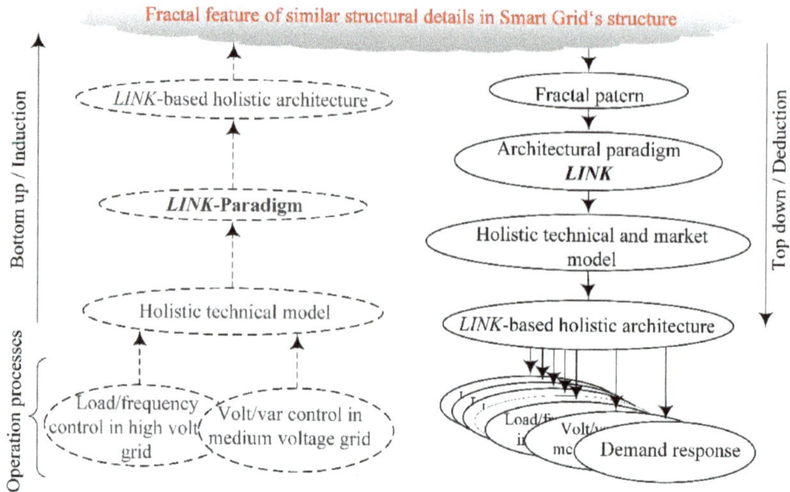

Figure 1. The methodology used to design the smart grids architecture and the market structure.

3. Results

3.1. Fractality in Smart Grids

Fractals are constructs characterized by a cascade of similar structural details that appear on all levels [40]. The cascade is never-ending in theoretical mathematical cases. However, in our study, we are restricted to an ending cascade starting with the highest voltage level and ending with the customer device level, as shown in Table 1. This means that we consider the power system holistically, regardless of the voltage levels, equipment sizes and technologies used.

Table 1. The fractal structure of smart grids based on different fractal levels.

Fractal Level	Grid			Active Power Appliances
	Wire	**Transformers**	**Reactive Power Devices**	**Appliances that Mainly Produce, Consume or Store Active Power**
Level 1. → HV_*EIA*	Very long	Very large	Very large	Very large
Level 2. → MV_*EIA*	Long	Large	Large	Large
Level 3. → LV_*EIA*	Short	Medium	Medium	Medium
Level 4. → CP_*EIA*	Very short	Small?	Small	Small
Level 5. → Dev_*EIA*	Tiny	Tiny	Tiny	Tiny

3.1.1. Fractal Pattern

The fractal of the entire power system (including the customer devices sporadically connected to the mains) is conceived as five fractal levels shown in Table 1. To effectively meet the ever-increasing demand for electricity, power engineers have designed and developed power systems based on different voltage levels. The electrical appliances connected in each of the three voltage levels, high, medium and low (HV, MV and LV, respectively) grids, are contained in fractal levels 1, 2 and 3, respectively.

The electrical appliances connected in the CP grid correspond to level 4. Level 5 takes into account all internal elements of the electrical devices that are connected by wires/conducting media. Power system appliances are categorized in two groups: The grid itself and active power appliances (APAs).

The grid is a construction that provides the connections amongst the points of power production and consumption. The relevant grid elements are wires, transformers and reactive power devices (RPDs) [56]. It is a fact that power, whether using alternating or direct current, needs wires to conduct it from the point of production to one of consumption. Thus, conductors are present in all fractal levels. In level 1, the high voltage level, they are called lines, while in the medium and low voltage levels, levels 2 and 3, they are more commonly referred to as feeders. High voltage lines are the longest, ranging from 100 to 2000 km. The world's longest power transmission line is the 600 kV direct current Rio Madeira transmission link in Brazil [57], with a length of 2385 km. Feeders in medium voltage grids are long, ranging from a few km to 70 km, while in low voltage grids they are short and range between several tens of meters and two kilometers. The CPs included in level 4 have short wires installed in building walls. They range between a few to several tens of meters.

Electrical devices connected to the customer plant grid have internal small/tiny wires that range from a few to several tens of centimeters. Thus, as the fractal level increases, the wire length decreases. The same trend also characterizes the wire diameters. The wire length and diameter repeatedly become smaller in the branches from level 1 upwards. Thus, wires are constructs that repeat themselves at progressively smaller scales.

The use of different voltage levels in the power system design dictated the introduction of transformers as connecting elements. Very large, large and medium sized transformers are used in HV, MV and LV (or fractal levels 1, 2 and 3), respectively. Transformers do not exist at the CP level, the gray elements in Table 1. Thus, the fractal structure presented in Table 1 shows a design anomaly in level 4. This anomaly may indicate further optimization potential of the power system structure. In the past, losses were one of the key optimization criteria in the power system design process. Only in the 1970s was it discovered that supplying a load using lower voltages reduces demand and energy consumption [58]. Since then, many studies based on conservation voltage reduction (CVR) have been conducted to assess the effectiveness of the implementation [59]. The current CVR implementation is based on controlling the voltage through distribution grids. Apart from the fact that its implementation in wide areas carries the risk of exposing some customers to unacceptable under-voltage conditions, the radial structures in distribution grids are subject to decreasing voltage profiles. The latter means that not all customers can be supplied with the lowest permitted voltage. This can be solved by installing 1:1 transformers with voltage control ability at the CP level, as indicated in the fractal structure given in Table 1. Their adjustment can guarantee the end customer's supply with the lowest possible voltage, thus ensuring maximum load reduction and energy savings. Additionally, this may lead to the liberalization of the voltage limits in super-ordinate grids, which in turn would make operation of the entire power system more efficient. However, basic economic studies are needed to substantiate this hypothesis. Tiny transformers exist in several customer devices (level 5). As in the wires case, the transformer size decreases with voltage level, meaning that it decreases with increasing fractal level. Transformers are thus constructs that repeat themselves at progressively smaller scales.

RPDs are mainly locally controlled (LC). They are widely used in the high and medium voltage grids to control the voltage. Very large and large sized RPDs are used in HV and MV (fractal levels 1 and 2), respectively. RPDs are not present in the LV level. The fractal structure presented in Table 1 thus shows a design anomaly in level 3, the gray element. In contrast, in level 4, the table is already filled, because with the integration of rooftop PVs, associated inverters are used to support the voltage control in low voltage grids by providing reactive power (almost purely inductive) [60]. Studies have shown that the use of customers' own devices to control the voltage on low voltage grids is technically not effective [61,62] and can lead to social problems [63]. The use of distribution system operators' (DSOs) reactive power sinks to control the voltage in LV grids is found to be more effective [61]. Thus, level 3 of the fractal structure table can be filled by DSOs and customer inverters should be used only to compensate for their reactive power requirements [62]. Tiny RPDs exist in several customer devices (level 5). The RPD size follows the same trend as discussed above; it decreases with increasing fractal level. RPDs are constructs that repeat themselves at progressively smaller scales.

APAs are appliances that mainly produce, consume or store active power (rotating machines, photo voltaic, batteries, etc.). Very large sized APAs present in hydro, nuclear, etc., power plants as well as in pumped hydro power plants (storage) are connected to HV grids (level 1). With the emergence of distributed energy resources, large, medium and small sized APAs are already connected in MV and LV grids and in CPs (levels 2, 3 and 4), respectively. Moreover, several devices such as computers have tiny APAs in the form of batteries (level 5). The column of APAs in Table 1 is completely filled. The APA size follows the same trend as discussed above; it decreases with increasing fractal level. Thus, APAs are constructs that repeat themselves at progressively smaller scales.

As a result, although very heterogeneous, the construction of smart grids is self-similar. Figure 2 shows its fractal pattern. Both groups of self-similar constructs (grid and APA) are figuratively wrapped in an ellipse, see Figure 2a.

By definition, the fractal pattern of the power system consists of electrical appliances (ElA) designed for a pre-defined level.

Each ElA-ellipse denotes a separate chain link. It includes the grid of a specific level and all APAs connected to it. Figure 2b illustrates ElA in different levels: HV-ElA, MV-ElA, LV-ElA, CP-ElA and Dev-ElA for levels 1, 2, 3, 4 and 5, respectively. Figure 2c magnifies the CP- and Dev-ElA patterns.

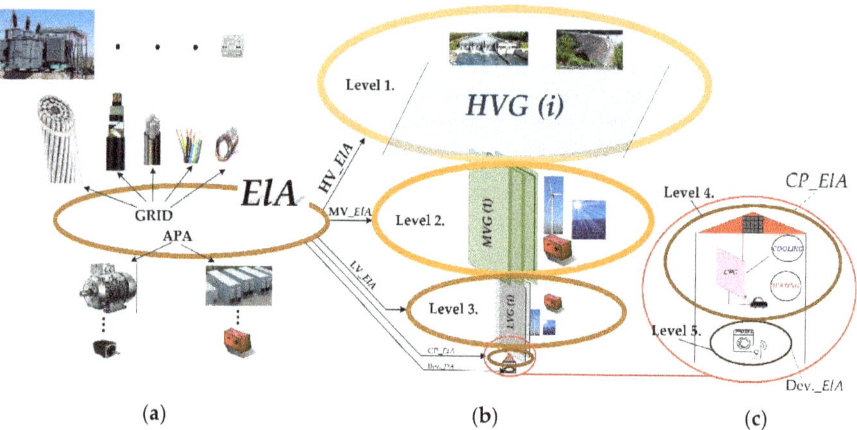

Figure 2. Link chain fractal of power systems: (**a**) The fractal pattern electrical appliances (ElA); (**b**) ElA in different fractal levels; (**c**) magnification of levels 4 and 5.

3.1.2. Fractal-Set of Smart Grids

From the bird's-eye view, all identified fractal levels are in the same territory, T. The overall pattern of level 1, HV-ElA, spreads over the entire T of a country or of a part of it, Figure 3. The patterns of the higher fractal levels propagate in smaller areas of the same territory T, repeating themselves many times. The repeating number of patterns in different fractal levels is derived from the relationship between different zones of power systems in real conditions and the number of appliances comprised in each of them. Table 2 shows a structural overview of the power system in real conditions and of its fractal.

Since the MV and LV grids have radial structures, the number of MV and LV zones are defined by the number of suppling substations (HV/MV) and distribution transformers (MV/LV), respectively.

A real European power system, operated by one transmission system operation (TSO), have about 300 supplying and 100,000 distribution substations to supply about 4 million customer plants.

Figure 3. The fractal of high voltage grids (HVGs).

Table 2. The fractal structure of smart grids based on different fractal levels.

Entire real Power System		Scaling Factors	Fractal of the Entire Power System	
Zone	**No. of Zones**		**No. of ElAs**	**Fractal Level**
HV	1	1	1	Level 1^{HV}
MV (Suppl. Substations)	300	50	6	Level 2^{MV}
LV (Distr. Substations)	100,000	5,555.56	18	Level 3^{LV}
CP (Customers)	4,000,000	55,555.56	72	Level 4^{CP}
Dev. (Elec. Devices)	40,000,000	55,555.56	720	Level 5^{Dev}

It is assumed that each distribution substation supplies on average about 40 customer plants [64]. Additionally, it is assumed that in each customer plant exists on average for ten electrical devices. Figure 4a shows a graphical overview of the structure of smart grids. The graph increases exponentially as the number of customer plants, 4 million, increases drastically compared to the number of the HV zones, that is one. Thus, the fractal set cannot be assembled using only one scaling factor, because the number of zones in different levels is very different. In Table 2 are shown the different scaling factors used to define the number of ElAs for each fractal level and the derived numbers of ElAs used. The scaling factor for level 1 is set to one, while for levels 2 and 3 they are set to 50 and 5,555.56, respectively. The same scaling factor of 55,555.56 is used for levels 4 and 5 because the number of zones included in them differs only by a factor of ten. Figure 4b shows a graphical overview of the structure of the fractal set. The graph behaves similarly to the one of the power system in real conditions, Figure 4a. Using this approach, each of the six MV-ElAs touches the single HV-ElA at one point and tangentially to each other, Figure 5.

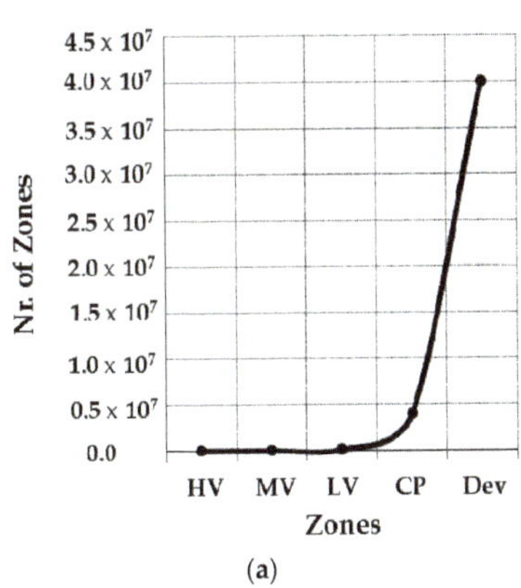

(a)

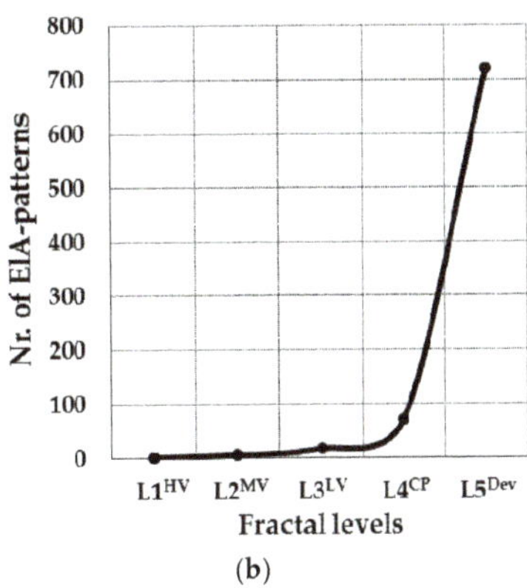

(b)

Figure 4. Overview of the structure of smart grids: (**a**) Real conditions; (**b**) fractal.

Figure 5. The fractal set of HV and medium voltage graphs (MVGs).

The latter symbolizes the tie switches (normally open points) connecting feeders of different MV-subsystems. For the 18 LV-ElAs is used the same principle as for MV-ElAs, Figure 6.

Figure 6. The fractal set of HV, MV and low voltage graphs (LVGs).

Four CP-ElAs are added into every LV-ElA, each touching only one point of the LV-ElA pattern (every customer plant has one electrical connection point with the LV grid and no connection with each other), Figure 7.

Figure 7. The fractal set of HV, MV and LVGs and customer plants (CP).

The same principle is used to add the ten Dev-ElAs into every CP-ElA, Figure 8. The fractal pattern ElA of different levels is assembled in a common fractal set; a link chain fractal limited to five interlinks.

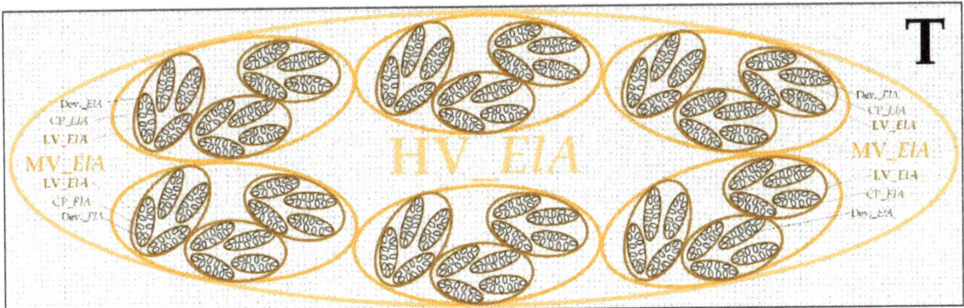

Figure 8. The fractal set of smart grids.

The significance of the fractal set of smart grids is verified by the calculation of the fractal dimension. It contains information about the geometrical structure at multiple scales reflecting the complexity of the fractal set: The greater the complexity, the larger the fractal dimension.

3.1.3. Fractal-Dimension

Fractals can be characterized by the fractal dimension *D*. For a fractal located in a bi-dimensional space, *D* is larger than 1 and less than 2,

$$1 < D < 2. \tag{1}$$

The larger *D*, the greater the fractal complexity.

D is calculated using the box-counting method, which is based on the counting of boxes (of side length (s) occupied by the fractal. The number of occupied boxes, *N*, increases by making the grid finer ($s \rightarrow 0$). D is estimated as the exponent of a power low as follows:

$$D = \lim_{s \to 0} \frac{\log N(s)}{\log\left(\frac{1}{s}\right)} \tag{2}$$

The software Fractalyse 2.4 [65] is used to calculate the fractal dimension of the entire power system, customer plants and electrical devices. The fractal dimension is obtained by using a logarithmic linear regression to fulfil the following objective function

$$\log N(s) = D \cdot \log\left(\frac{1}{s}\right) + C, \tag{3}$$

where C is the equation constant.

Figure 9 shows the separate fractal sets of smart grids in different levels. While the logarithm diagram for measuring D of each level is shown in Figure 10. The segments' slope increases steadily with increasing fractal level. Figure 11 shows the fractal dimension D for different separate fractal levels. D increases from 1184 for level 1 (HV) to 1272 and 1308 for the levels 2 and 3 (MV and LV), respectively. The D-increase from level 1 to level 2 is significant ($\Delta D^{L1 \rightarrow L2} = 0.088$), which means that the complexity of smart grids at MV level increases dramatically compared to the HV level. Another significant D-increase, even bigger than the previous one, is observed between levels 3 (LV) and 4 (CP), ($\Delta D^{L3 \rightarrow L4} = 0.108$). The complexity of smart grids at the CP level increases considerably compared to the LV level. The slope of the trend line of levels 4 and 5 (dashed line) is clearly bigger than in the case of the trend line for levels 2 and 3 (dotted line), 0.156 and 0.0347, respectively.

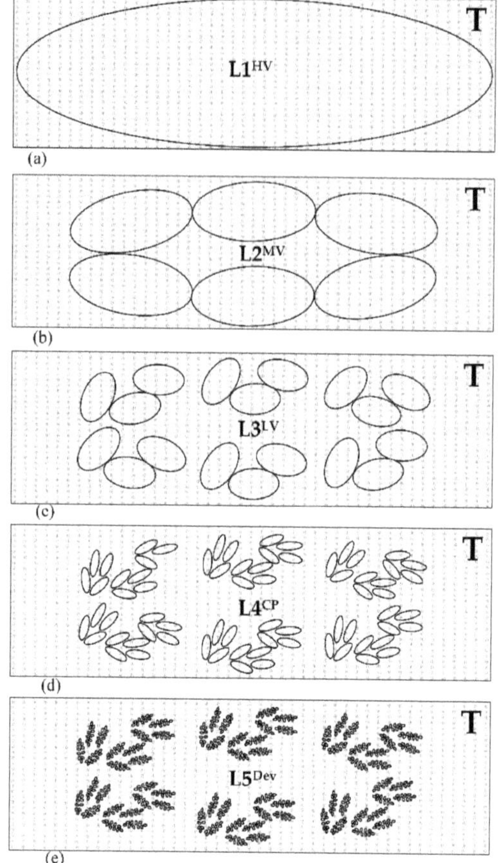

Figure 9. Smart grid's separate fractal sets in different levels: (**a**) Level 1; (**b**) level 2; (**c**) level 3; (**d**) level 4; (**e**) level 5.

The increase in complexity at the level of electrical device compared to the customer plant is stronger than that between the LV and MV levels.

Usually, the resources required to develop and purchase the different parts of a complex system are proportional to their complexity. The fractal dimension analysis indicates that, to realize smart grids, the highest global resources should be provided for the development and purchase of electrical devices, followed by a continuous reduction in resources for CPs, LV and MV up to the lowest resource allocation for the HV level.

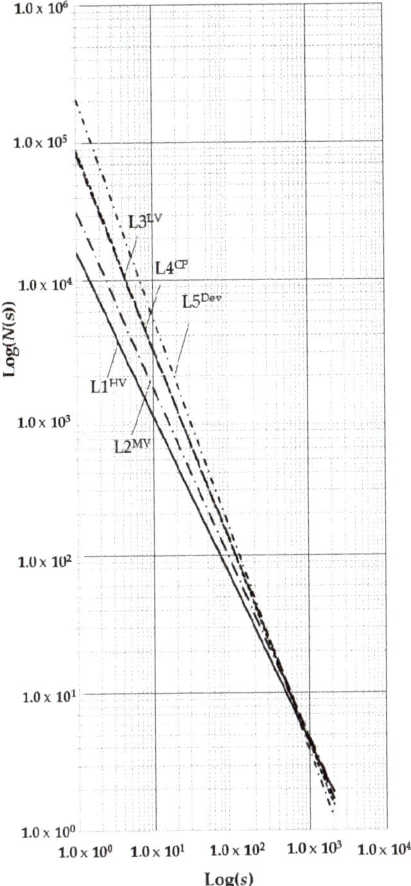

Figure 10. Logarithmic diagram for measuring the fractal dimension of different separate fractal levels.

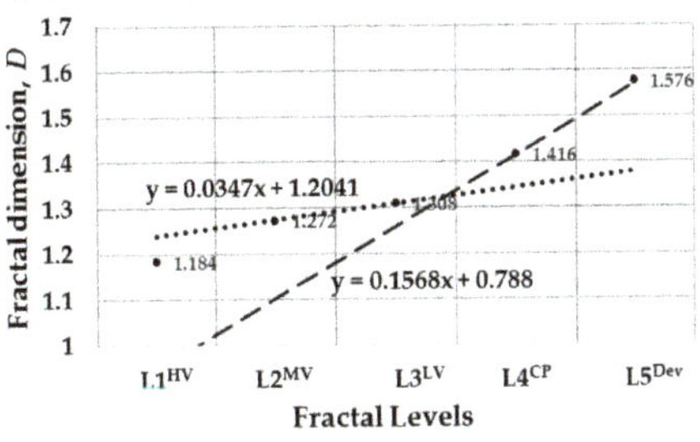

Figure 11. The fractal dimension D for different separate fractal levels.

3.2. *LINK-Paradigm and the Associated Holistic Model*

The realization of smart grids is related to a growing demand for more sensors, more communication, more computation and more control [66]; these represent an ever-growing cornucopia of new technologies [67,68]. Therefore, the ElA fractal pattern is combined with control schemes and interfaces to create the smart grid paradigm, Figure 12.

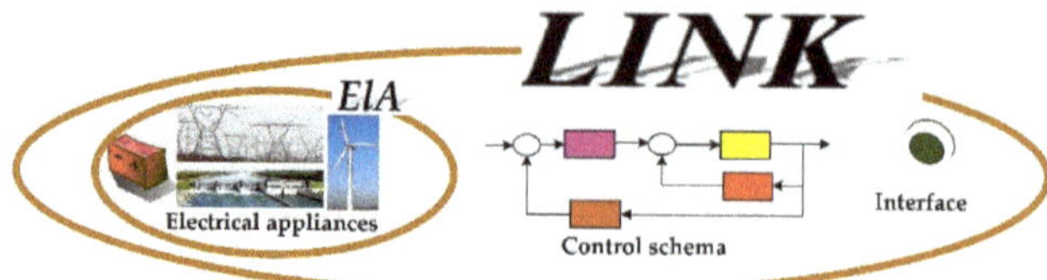

Figure 12. Overview of the smart grid paradigm.

By definition, the *LINK*-paradigm is a set of one or more ElAs, i.e., a grid part, storage device or a producer device, the controlling schema and the interface.

The *LINK*-paradigm is used as an instrument to design a *LINK*-based holistic architecture. It facilitates modelling of the entire power system from high to low voltage levels, including CPs and includes the description of all power system operation processes such as load-generation balance, voltage assessment, dynamic security, price and emergency driven demand response, etc. [69]. The *LINK*-paradigm is the fundament of the holistic, technical and market-related model of smart power systems with large distributed energy resource (DER) shares. Figure 13a shows the technical holistic model (the "energy supply chain net"), which extends to fractal level 4. It illustrates the links' compositions and their relative position in space, both horizontally and vertically. In the horizontal axis, the interconnected high voltage grids (HVGs) are arranged. They are actually owned and operated by TSOs. The medium and low voltage grids (MVGs and LVGs, respectively) and the customer plant grids (CPGs) are arranged on the vertical axis. MVGs and LVGs are actually owned and operated by the DSOs, while CPGs are operated by customers.

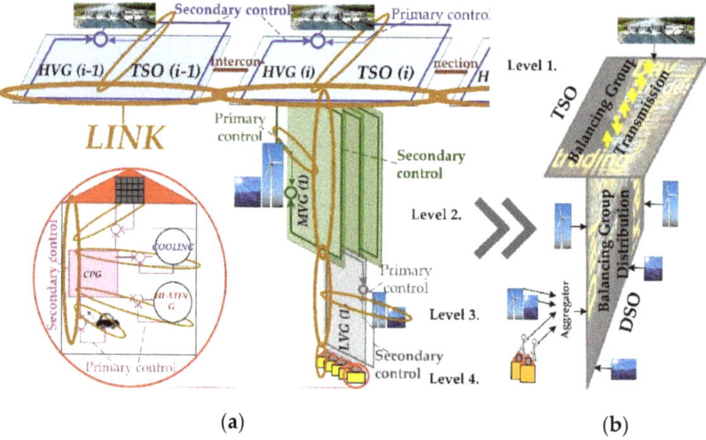

Figure 13. Overview of the holistic models: (**a**) Technical, the "energy supply chain net"; (**b**) market.

By definition, an "energy supply chain net" is a set of automated power grids intended for chain links (abbreviated as links), which fit into one another to establish a flexible and reliable electrical connection. Each individual link or link bundle operates autonomously and has contractual arrangements with other relevant boundary links or link bundles [70].

The holistic model associated with the energy market is derived from the technical holistic model, the "energy supply chain net", as shown in Figure 13b. The whole energy market consists of coupled market areas (balancing groups) at the horizontal and vertical axes. TSOs operate on the horizontal axis of the holistic market model, while DSOs operate on the vertical. Based on this model, not only TSOs but also DSOs will communicate directly with the whole market to ensure a congestion-free distribution grid operation and to take over the task of load-production balance. The owner of the distributed energy resources as well as the prosumers (producers and consumers of electricity) may participate directly in the market or may do so via aggregators or local energy communities (LECs) [71]. The creation of the local retail markets (LRM) attracts the demand response bids and stimulates investment in the LEC areas (see Section 3.3.2).

3.3. LINK-Based Holistic Architecture

When the structure of electricity supply changes radically because of many DG units, each with the possibility of interfering with the system operation at all voltage levels, then a new architecture is needed to utilize this added flexibility so that power system operation can remain reliable. The new operational architecture of power systems should guarantee that it performs as expected by unifying their whole structure and by systematizing the execution of operational tasks. The basis for the design of the *LINK*-based holistic architecture is in its definition [39,72]:

Definition: A holistic power system architecture is an architecture in which all relevant components of the power system are merged into one single structure. These components could consist of the following: Electricity producer (regardless of technology or size, e.g., large power plants, DGs, etc.); electricity storage (regardless of technology or size, e.g., pumped power plants, batteries, etc.); electricity grid (regardless of voltage level, e.g., high, medium and low voltage grids); customer plants and electricity market.

Fleshing out the holistic model into a schematic holistic architecture requires specification of the main independent architecture elements, Figure 14. There exist three independent main power system components, i.e., producer, storage and grid, which are the three basic elements of the architecture and create the basis for the definition of the architecture elements. They are defined as follows:

1. Producer link is a composition of an electricity production facility, be it a generator, photovoltaic, etc., its primary control (PC) or the producer interface. Primary control refers to control actions that are done locally (device level) based on predefined set-points. The actual values are measured locally and deviations from the set-points results in a signal that influences the valves, excitation current, transformer steps, etc. in a primary-controlled power plant, transformer, etc., such that the desired power is delivered or the desired voltage is reached;
2. Storage link is a composition of a storage facility, be it the generator of a pump power plant, batteries, etc., its PC and the storage interface;
3. Grid link is a composition of a grid part, called a link grid, with the corresponding secondary control (SC) and link interfaces. SC refers to control actions that are calculated based on the grid link control area. It fulfils a predefined objective function by respecting the dynamic grid constraints on the grid link boundaries and the static constraints of electrical appliances (PQ diagrams of generators, transformer rating, etc.). Dynamic grid constraints are the reactive and active power exchange on the grid link boundaries that are agreed from the corresponding grid link operators [73]. The grid link contains SCs for both major entities of power systems—frequency and voltage. The SC algorithm fulfils technical issues and calculates the set points by respecting the dynamic constraints, which are necessary for stable and reliable operation. The link grid size is variable and is defined from the area where the secondary control is set up. Thus, a link grid may apply to a customer plant or even to a large high voltage grid area.

To overcome data privacy and cyber security challenges, a distributed *LINK*-based holistic architecture is chosen [38]. Its key principle is to prohibit access to all resources by default, allowing access only to a minimum of data. Each link has its own operator. Based on the architecture elements, operators can be classified into three types:

(1) The producer link operator operates each power plant regardless of technology and size (excluding very small power plants, for example PVs, installed on the customer side);
(2) The storage link operator operates each storage regardless of technology and size (excluding very small storage, for example batteries, installed on the customer side);

(3) The grid link operator operates the grid regardless of voltage level (excluding the customer grid). Customers themselves are responsible for the operation of all their home elements.

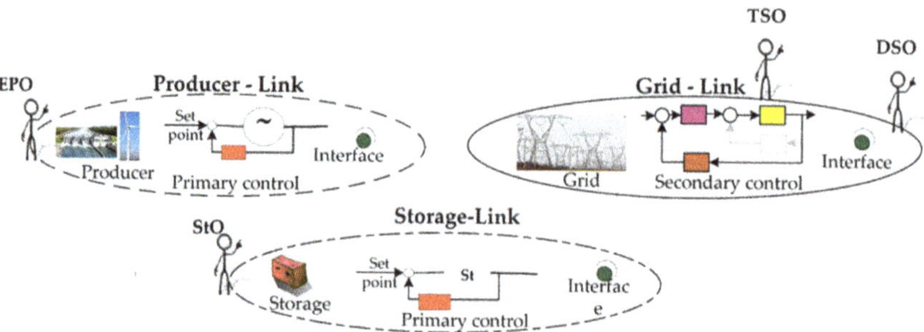

Figure 14. Main elements of the *LINK*-based holistic architecture and the corresponding operators.

The technical/functional stage includes the conjunction of all three architecture elements in the four fractal levels (Figure 15a)) and a standardized architecture structure encompasses the four fractal levels. The different links communicate with each other via well-defined technical interfaces [38] "T". This is a detailed architecture level facilitating all technical/functional processes, which are needed to reliably and economically operate the decentralized power system. All processes can be described by using the unified modelling language (UML) diagrams (see Section 3.3.3). Additionally, it enables the step into the level of management systems, i.e., the application level, to develop concrete applications, (see details in the "Methods" section). The standardized structure of the technical/functional architecture allows the transition to the generalized architecture level (Figure 15b), where the four fractal levels are represented very compactly. The generalized architecture is the core of the *LINK*-based holistic architecture (Figure 15c). The market surrounds it and communicates with it through the market interfaces, "M", by exchanging aggregated meter readings, external schedules, etc. [39]. At the holistic level, the grid links of customer plants are removed from the generalized presentation because they are too small to participate directly in the whole market. They may participate in the common market through aggregators or through energy communities [71]. For the sake of privacy and cyber security, the technical interfaces "T" are designed apart from the market interfaces "M". The new designed holistic architecture facilitates two operating modes: (1) Autonomous—each individual link or link bundle operates independently by respecting the contractual arrangements with other relevant boundary links or link bundles. All links are connected together creating a large power system; (2) autarkic or self-sufficient—an optional operating mode, which may be applied in any link bundle that consists of at least one grid link and one producer link or storage link as long as it is self-sufficient and self-sustaining without any dependency on electricity imports. Restoration—an option of the autarkic operating mode, which may be applied after a blackout during the restoration process to supply electricity to at least the communication appliances. In order to successfully switch the operating mode from autonomous to autarkic, a familiar resynchronization process should be established, where each grid link has a secondary control on active and reactive power that supports the synchronization process. Depending on the properties of the links, the resynchronization with other links may be automatic or manual. However, resynchronization philosophies should be initially investigated to determine the most appropriate approach.

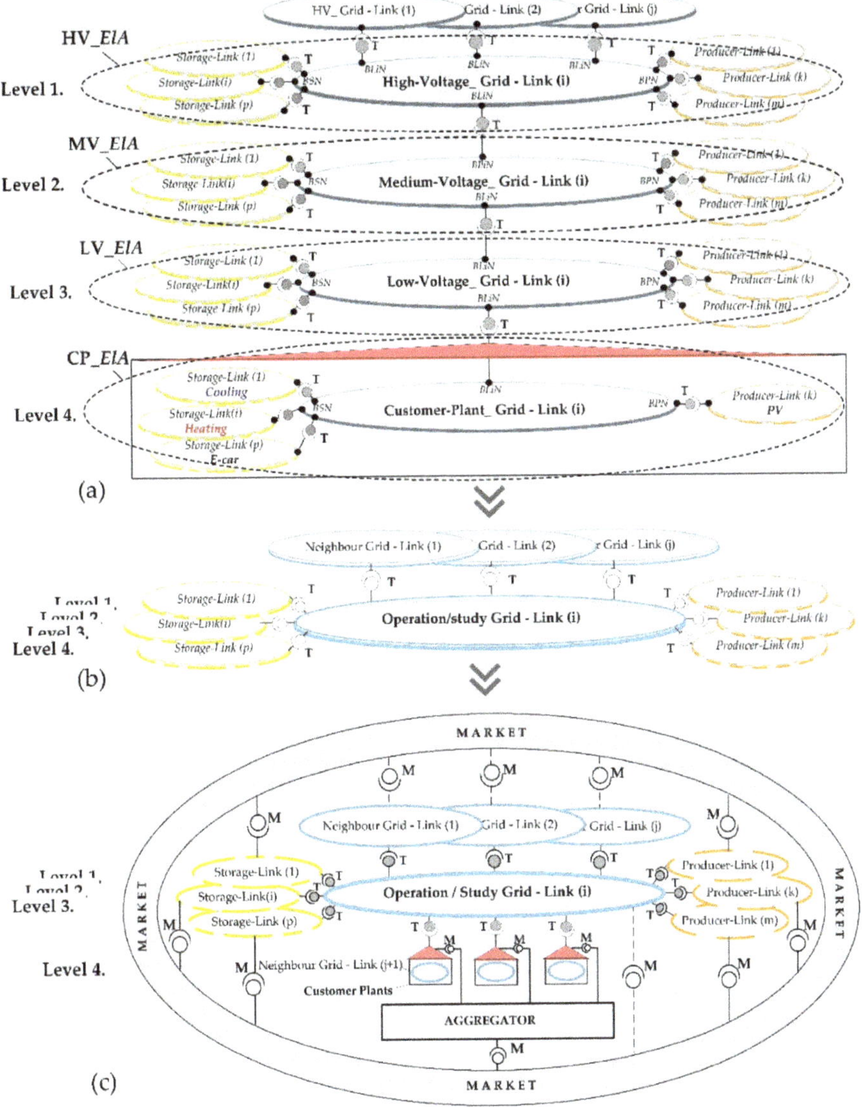

Figure 15. Different stages of the *LINK*-based architecture: (**a**) Technical/functional, (**b**) generalized and (**c**) holistic.

Operators are responsible for safe, reliable and efficient operation of the power grid at all times. The *LINK*-based holistic architecture supports them to achieve these goals, facilitating the execution of a set of functions and tasks that are encapsulated in various technical/functional operational processes, such as: Monitoring, generation-load balance [38], voltage assessment and var management [67], static and dynamic security [38], emergency [38], price driven demand response (see Section 3.3.3), etc. Some applications in the context of the holistic architecture are given in the following.

3.3.1. Compact Representation of the Control-Chain Strategy

The standardized architecture of *LINK*-based architecture enables a compact and clear representation of the control strategy, which is designed as a chain net. According to Section 3.2, the latter is conceived on the X- and Y-axes. The different units of secondary control form a chain, which is expressed below in terms of volt/var control (VVC) chain.

The VVC chain in the X-axis extends to very HV and HV grids and is presented by the following generalized equation:

$$VVC^{X-axis} = \left\{ VV\boldsymbol{SC}^{HV}\left(Volt\boldsymbol{PC}^{HV}_{OLTC}, var(\boldsymbol{PCorLC})^{HV}_{RD}, var\boldsymbol{PC}^{HV}_{G}, var\boldsymbol{Ngb}GridLink^{HV}_{HV,MV,LV} \right) \right\} \tag{4}$$

where $VV\boldsymbol{SC}^{HV}$ calculates:

- the voltage set-point $Volt\boldsymbol{PC}^{HV}_{OLTC}$ for each transformer included on the HV link grid that has an on-load tap changer (OLTC);
- the var set point $var\boldsymbol{PC}^{HV}_{RD}$ or the switch position of $var\boldsymbol{LC}^{HV}_{RD}$ for all RDs connected on the HV link grid;
- the var set point $var\boldsymbol{PC}^{HV}_{G}$ for all generators connected on the HV link grid;
- the var set point $var\boldsymbol{Ngb}GridLink^{MV}_{MV,LV}$ for all neighbour HV, MV or LV grid links when available.

While, the VVC chain in the Y-axis extends to HV, MV and LV grids and CP. It is presented by the following generalized equation:

$$\begin{aligned} VVC^{Y-axis} = \Big\{ & VV\boldsymbol{SC}^{HV}\left(Volt\boldsymbol{PC}^{HV}_{OLTC}, var(\boldsymbol{PCorLC})^{HV}_{RD}, var\boldsymbol{PC}^{HV}_{G}, var\boldsymbol{Ngb}GridLink^{HV}_{HV,MV,LV} \right), \\ & VV\boldsymbol{SC}^{MV}\left(Volt\boldsymbol{PC}^{MV}_{OLTC}, var(\boldsymbol{PCorLC})^{MV}_{RD}, var\boldsymbol{PC}^{MV}_{DG}, var\boldsymbol{Ngb}GridLink^{MV}_{MV,LV,LV}, var\boldsymbol{Cns}^{MV}_{HV} \right), \\ & VV\boldsymbol{SC}^{LV}\left(var\boldsymbol{PC}^{LV}_{RD}, var(\boldsymbol{PCorLC})^{MV}_{RD}, var\boldsymbol{PC}^{LV}_{DG}, var\boldsymbol{Ngb}GridLink^{LV}_{CP}, var\boldsymbol{Cns}^{LV}_{MV} \right), \\ & VV\boldsymbol{SC}^{CP}\left(var\boldsymbol{PC}^{CP}_{inv}, var\boldsymbol{Cns}^{CP}_{LV} \right) \Big\}, \end{aligned} \tag{5}$$

where $VV\boldsymbol{SC}^{MV}$ calculates:

- the voltage set-point $Volt\boldsymbol{PC}^{MV}_{OLTC}$ for the supplying transformer and for other transformers included on the MV link grid (e.g., 34.5 kV/11 kV, etc.) that have OLTC;
- the var set point $var\boldsymbol{PC}^{MV}_{RD}$ or the switch position of $var\boldsymbol{LC}^{MV}_{RD}$ for all RDs connected on the MV link grid;
- the var set point $var\boldsymbol{PC}^{MV}_{DG}$ for all DGs connected on the MV link grid;
- the var set point $var\boldsymbol{Ngb}GridLink^{MV}_{MV,LV}$ for all neighbour MV or LV grid links when required, while respecting the var constraint on the border with HV $var\boldsymbol{Cns}^{MV}_{HV}$.

$VV\boldsymbol{SC}^{LV}$ calculates:

- the var set point $var\boldsymbol{PC}^{LV}_{RD}$ for all RDs connected on LV link grid;
- the var set point $var\boldsymbol{PC}^{LV}_{DG}$ for all DGs connected on LV link grid;
- the var set point $var\boldsymbol{Ngb}GridLink^{LV}_{CP}$ for all neighbour CP grid links when required, while respecting the var constraint on the border with MV $varCns^{LV}_{MV}$.

$VV\boldsymbol{SC}^{CP}$ calculates:

- the var set point $var\boldsymbol{PC}^{CP}_{inv}$ of the inverter connected on CP link grid; while respecting the var constraint on the border with LV $var\boldsymbol{Cns}^{CP}_{LV}$ Some practical applications of Equation (5) are shown in Section 3.3.4 and in reference [74].

3.3.2. Smart Grid and Market Harmonized Structures

The large-scale DER integration increases the efficiency of the power grid and helps with its decarburization. In addition, their existence contributes to the LRM design, enabling their market-based control, which in turn stimulates investment in the LEC areas.

The market-based control must take into account the technical behaviour of the smart grids and support their safe reliable and resilient operation, while at the same time attracting the demand response bids. This is illustrated below by describing the harmonization and coordination of the market structure with the grid link arrangement, Figure 16. The implementation of LRMs requires the creation of balancing group areas (BGA) within the whole market, the day-a-head market that harmonizes with the grid links as follows:

- BGA is a geographical area consisting of one or more grid links with common market rules and has the same price for imbalance in the day-a-head market. In general, a TSO grid includes one grid link, while the grid of a DSO may include several grid links. Additionally, all grid links included in one BGA should be operated by one DSO (not by several DSOs).
- The geographic boundaries of BGA may vary considerably and are defined by the external boundaries of grid links contained in the BGA. Grid links may have two types of boundaries: external and internal. The external boundaries exist between different links, which have different owners or operators i.e., between TSO and DSO. They are subject to the data security and privacy because of the data exchanges between two different companies. Internal interfaces are set between different links, which have the same owner or operator, e.g., the same DSO. They are subject only to the data security [73].
- All electricity producers, storages and consumer, that are electrically connected to the grid links belonging to the BGA participate in the same balancing group.
- The grid link operators (GLO may be a TSO, DSO or LEC) should act as neutral market facilitators having the responsibility to control and independently balance power flow fluctuations in their own area and for the secure and reliable operation of the grid. Moreover, each GLO provides ancillary services to the neighbor grid link areas to ensure the reliability of the electricity supply.

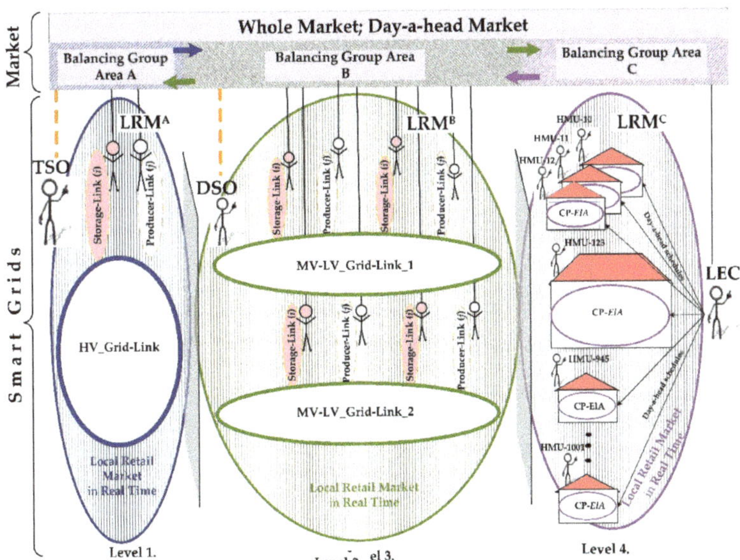

Figure 16. Harmonization of the market structure with the grid link arrangement.

The electricity market is thus similar to the physical reality of power flows, as neighboring grids links are physically connected anyway and electricity always takes the shortest route from producer to consumer—across grid links and market area boundaries. The rules and methods for LRM operation should be in line with national policies and are therefore the subject of other studies.

The harmonized electricity market structure facilitates the introduction of the demand response process in large scale as follows.

3.3.3. Demand Response Process

This is illustrated below by describing the process of price driven demand response. The activation of residential, commercial and small business sectors, which join the real-time pricing demand response through already concluded contracts, may be triggered at any time. Their degree of participation in the demand response process may be different depending on the time of the day, duration interval, price value, etc. In the case of a surplus of electricity in the market, the electricity price decreases. All market participants and market operators will be notified to allow them the possibility to act on time. Figure 17 shows the information flow during price driven demand response. This enables residential, commercial and small business sectors to perceive transparent energy prices and to contribute in the reliable and efficient operation of electric power systems.

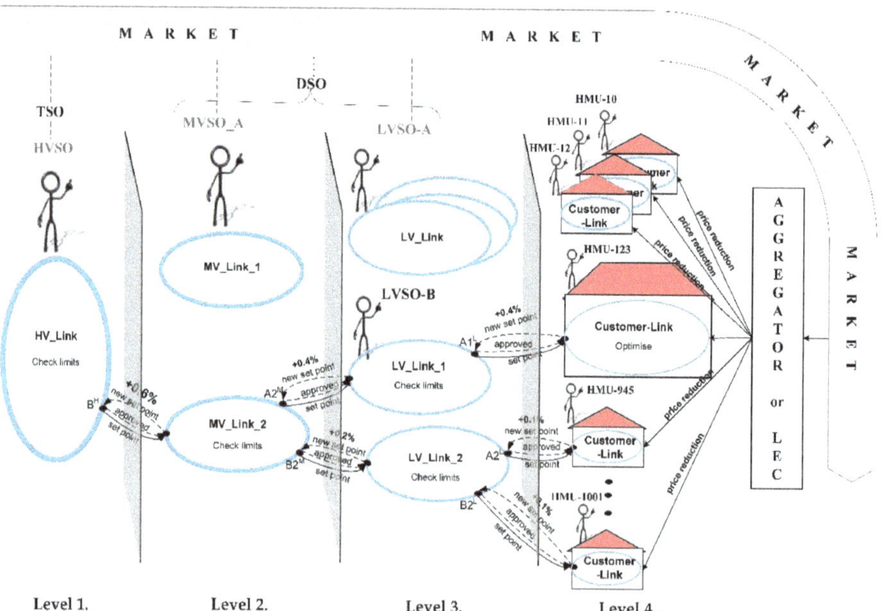

Figure 17. Unified modelling language (UML) diagram of the process price driven demand response.

3.3.4. Conservation Voltage Reduction in MV Level

The proof of the concept is performed in the frame of the industrial research project ZUQDE (central volt/var control in presence of DGs), Salzburg, Austria, based on one of the major entities of power systems, the voltage [75,76]. The scope was fractal level 2 (MV), where the secondary control link was realized by means of the volt/var control, Figure 18.

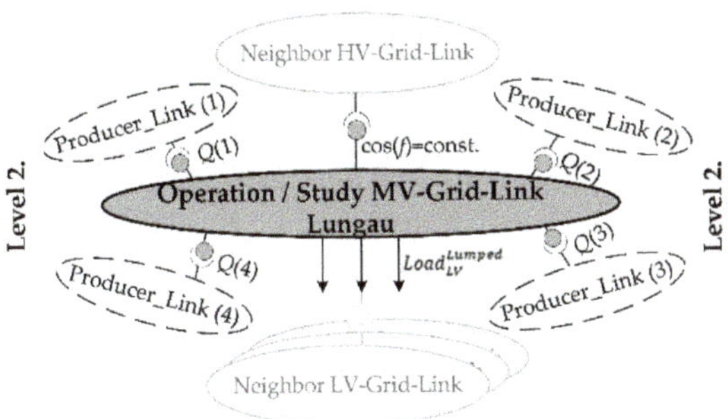

Figure 18. The implemented technical/functional *LINK*-based architecture in the ZUQDE project.

The applied algorithm calculated the set points by respecting the constraint, which was set to the HV/MV transformers by means of a constant cosΦ. The set points, reactive power Q and voltage, were sent to all four "run of river" distributed generators (DG) and to the feeder head bus bar, respectively, Figure 19. All relevant generators were upgraded along with the primary control, thus building up the producer component. All distributed transformers were modelled loads. As a result, the voltage in the Lungau region was automatically controlled and the grid was dynamically optimized in real time for more than one year.

$$VVC^{Y-axis} = \left\{ VVSC^{MV}\left(VoltPC^{MV}_{OLTC}, varPC^{MV}_{DG},\ \cos(\varphi) Cns^{MV}_{HV} \right) \right\} \tag{6}$$

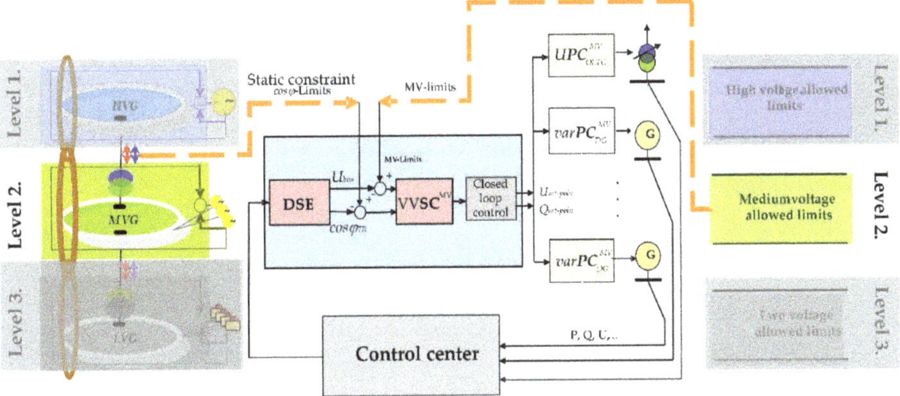

Figure 19. The realized volt/var control scheme in management system level in ZUQDE project.

The voltages, power and current during the CVR switching process are shown in Figure 20. The snapshot is made on the SCADA system of the control center. The curves of voltage in one of the 30 kV bus bars and of the active, measured on the supplying transformer level, are highlighted.

Significant in the consideration are the curves shown in blue, green and yellow. The blue curve shows the voltage set point on the 30 kV side of the supplying transformer. At 10:18 a.m., the VVC set in the CVR mode calculated new set points for the 30 kV bus voltage calculated new set points for the 30 kV bus voltage and the reactive power of DGs. The voltage set point changed from 31.7 kV to 30.5 kV. This corresponds to an adjustment of the on-load tap changer by two steps. The green curve shows the measured voltage on the 30 kV bus bar. Due to the reaction time of the OLTC, the new voltage set point was reached after about one minute. The yellow curve shows the active power, which decreases from 19.9 MW to 19.0 MW. This corresponds to a reduction of the active power by 4.7%. The light blue or turquoise curve shows that the voltage reduction results in a slight increase of the current.

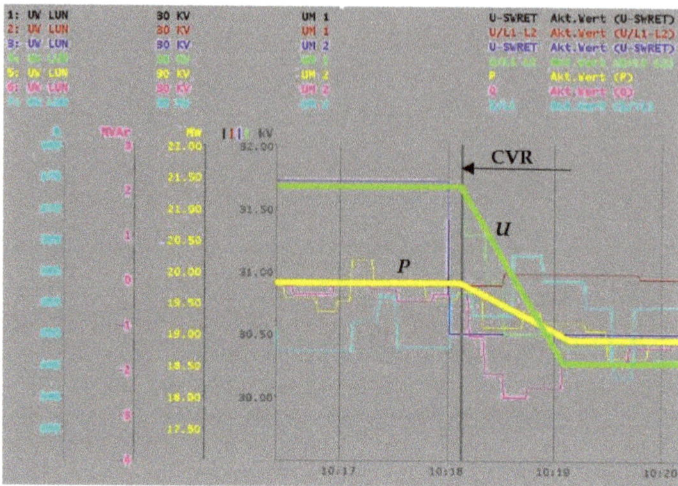

Figure 20. Voltage and active power course during the execution of the conservation voltage reduction (CVR) process in the ZUQDE project [76].

4. Conclusions

The definition of the architectural paradigm *LINK* is finally validated by the fractal analysis: It consists of unique and independent elements that avoid misinterpretation or the need for any changes in its definition. Its use allows the convergence of the results of various scientific works and the implementation of smart grids on a large scale.

The fractal analysis is a suitable instrument for characterizing the heterogeneity of the smart grid structure, as it is capable of characterizing the spatial differences within the smart grid. Based on fractal analysis, two fractal anomalies are identified in the existing power system structure. Firstly, 1:1 transformers with voltage control capability are missing in fractal level 4. Their presence could contribute to increased energy savings, which is a key objective under actual climatic conditions. However, fundamental studies are needed to assess this change in the current power grid structure. Secondly, reactive devices are missing in fractal level 3. Studies have shown that eliminating this anomaly produces essential benefits. The fractal dimension analysis indicates that to realize smart grids, the highest global resources should be provided for the development and purchase of electrical devices, followed by a continuous reduction in resources for CPs, LV and MV up to the lowest resource allocation for the HV level.

The *LINK*-based holistic architecture reflects the principles of the fractal. Repeating the same structures in ever smaller dimensions greatly improves the practical relevance of the *LINK* solution. The compact and clear representation of the control strategy, which is designed as a chain net, is straightforward und understandable for every electrical engineer. The *LINK*-based holistic architecture considers the entire power system from high, medium and low voltage levels, including customer plants and the market. It facilitates the description of all power system operation processes such as load-generation balance, voltage assessment, dynamic security processes, etc. It enables the large-scale DER implementation, respecting data privacy and minimizing cyber security risks. The new market structure harmonizes with the grid link arrangement, facilitating the demand response process. The CVR process is also easy to implement.

The realization of significant processes required for the operation of smart grids, such as demand response and load/generation balance need further investigation to show the effectiveness using the proposed architectural *LINK*-paradigm.

Funding: This research received no external funding.

Acknowledgments: Open Access Funding by TU Wien.

Conflicts of Interest: The author declare no conflict of interest.

References

1. Feynman, R.P. *The Character of Physical Law*; Modern Library: New York, NY, USA, 1994.
2. Maier, M.W.; Rechtin, E. *The Art of Systems Architecting*; CRC Press, Taylor & Francis Group: Boca Raton, FL, USA, 2009; ISBN 9781420079135.
3. Wolak, F.A. Diagnosing the California Electricity Crisis. *Electr. J.* **2003**, *16*, 11–37. [CrossRef]
4. Burr, M.T. Reliability demands will drive automation investments. *Fortnightly Magazine.* 1 November 2003, pp. 1–4. Available online: https://www.fortnightly.com/fortnightly/2003/11/technology-corridor (accessed on 3 May 2016).
5. Owens, B. *The Rice of Distributed Power*; General Electric Company: New York, NY, USA, 2014; Volume 47, Available online: https://www.ge.com/sites/default/files/2014%2002%20Rise%20of%20Distributed%20Power.pdf (accessed on 20 November 2018).
6. Myrda, P. *Smart Grid Enabled Asset Management*; Report 1017828; EPRI: Paolo Alto, CA, USA, 2009.
7. Hendschin, E.; Uphaus, F.; Wiesner, T.H. The Integrated Service Network as a Vision of the Future Distribution Systems. In *Proceedings of the International Symposium on Distributed Generation: Power System and Market Aspects*; Royal Institute of Technology: Stockholm, Sweden, 2001.
8. Stothert, A.; Fritz, O.; Sutter, M. Optimal Operation of a Virtual Utility. In *Proceedings of the International Symposium on Distributed Generation: Power System and Market Aspects*; Royal Institute of Technology: Stockholm, Sweden, 2001.
9. Dielmann, K.; Veiden, A. Virtual power plants (VPP)—A new perspective for energy generation? In Proceedings of the 9th International Scientific and Practical Conference of Students, Post-Graduates Modern Techniques and Technologies, Tomsk, Russia, 7–11 April 2003; pp. 18–20.
10. Zhang, G.; Jiang, C.; Wang, X. Comprehensive review on structure and operation of virtual power plant in electrical system. *IET Gener. Transm. Distrib.* **2019**, *13*, 145–156. [CrossRef]
11. Bignucolo, F.; Caldon, R.; Prandoni, V.; Spelta, S.; Vezzola, M. The Voltage Control on MV Distribution Networks with Aggregated DG Units (VPP). In Proceedings of the 41st International Universities Power Engineering Conference, Newcastle-upon-Tyne, UK, 6–8 September 2006; Volume 1, pp. 187–192. [CrossRef]
12. Pudjianto, D.; Ramsay, C.; Strbac, G. Virtual power plant and system integration of distributed energy resources. *IET Renew. Power Gener.* **2007**, *1*, 10–16. [CrossRef]
13. Morais, H.; Kadar, P.; Cardoso, M.; Vale, Z.A.; Khodr, H. VPP operating in the isolated grid. In Proceedings of the IEEE Power and Energy Society General Meeting—Conversion and Delivery of Electrical Energy in the 21st Century, Pittsburgh, PA, USA, 20–24 July 2008; pp. 1–6.
14. Saboori, H.; Mohammadi, M.; Taghe, R. Virtual Power Plant (VPP), definition, Concept, Components and Types. In Proceeding of the Asia-Pacific Power and Energy Engineering Conference (APPEEC), Wuhan, China, 25–28 March 2011.
15. Lund, P. The Danish cell project—Part 1: Background and general approach. In Proceeding of the IEEE Power Engineering Society General Meeting, Tampa, FL, USA, 24–28 June 2007; pp. 24–28.
16. Cherian, S.; Keogh, B.; Pacific, O. Dynamic distributed power grid control system. WO Patent 2012/008979, 19 January 2012.
17. Mazzi, N.; Trivella, A.; Morales, J.M. Enabling Active/Passive Electricity Trading in Dual-Price Balancing markets. *IEEE Trans. Power Syst.* **2019**, *34*, 1980–1990. [CrossRef]
18. Candra, D.I.; Hartmann, K.; Nelles, M. Economic Optimal Implementation of Virtual Power Plants in the German Power market. *Energies* **2018**, *11*, 2365. [CrossRef]
19. Lasseter, B. Microgrids [distributed power generation]. In Proceedings of the IEEE Power Engineering Society Winter Meeting, Columbus, OH, USA, 28 January–1 February 2001; Volume 1, pp. 146–149.

20. Lasseter, B. Microgrids. In Proceedings of the IEEE Power Engineering Society Winter Meeting, New York, NY, USA, 27–31 January 2002; Volume 1, pp. 305–308.
21. Soshinskaya, M.; Crijns-Graus, W.H.; Guerrero, J.M.; Vasquez, J.C. Microgrids: Experiences, barriers and success factors. *Renew. Sustain. Energy Rev.* **2014**, *40*, 659–672. [CrossRef]
22. Olivares, D.E.; Mehrizi-Sani, A.; Etemadi, A.H.; Cañizares, C.A.; Iravani, R.; Kazerani, M.; Hajimiragha, A.H.; Gomis-Bellmunt, O.; Saeedifard, M.; Palma-Behnke, R.; et al. Trends in microgrid control. *IEEE Trans. Smart Grid* **2014**, *5*, 1905–1919. [CrossRef]
23. IEEE Task Force on Load Representation for Dynamic Performance. Load representation for dynamic performance analyses. *IEEE Trans. Power Syst.* **1993**, *8*, 472–482. [CrossRef]
24. Stadler, M.; Naslé, A. Planning and implementation of bankable microgrids. *Electr. J.* **2019**, *32*, 24–29. [CrossRef]
25. Kleineidam, G.; Krasser, M.; Reischböck, M. The cellular approach: Smart energy re-gion Wunsiedel. Testbed for smart grid, smart metering and smart home solutions. *Electr. Eng.* **2016**, *4*, 335–340. [CrossRef]
26. Kießling, A. *Beiträge von moma zur Transformation des Energiesystems für Nachhaltigkeit, Beteiligung, Regionalität und Verbundheit. Modellstadt Mannheim (moma). Projekt Abschlussbericht*; MVV Energie: Mannheim, Germany, 2013.
27. Benz, T.; Dickert, J.; Erbert, M.; Erdmann, N.; Johae, C.; Katzenbach, B.; Glaunsinger, W. The Cellular Approach. VDE-Study. 2015. Available online: https://shop.vde.com/de/copy-of-vde-studie-der-zellulare-ansatz (accessed on 16 February 2017).
28. Morch, A.Z.; Jakobsen, S.H.; Visscher, K.; Marinelli, M. Future control architecture and emerging observability needs. In Proceedings of the IEEE 5th International Conference on Power Engineering, Energy and Electrical Drives, Riga, Latvia, 11–13 May 2015; pp. 234–238.
29. Syed, M.H.; Burt, G.M.; Kok, J.K. Ancillary Service Provision by Demand Side Management: A Real-Time Power Hardware-in-the loop Co-simulation Demonstration. In Proceedings of the International Symposium on Smart Electric Distribution Systems and Technologies (EDST), Vienna, Austria, 8–11 September 2015. [CrossRef]
30. D'hulst, R.; Fernandez, J.M.; Rikos, E.; Kolodziej, D.; Heussen, K.; Geibelk, D.; Temiz, A.; Caerts, C. Voltage and frequency control for future power systems: The ELECTRA IRP proposal. In Proceedings of the International Symposium on Smart Electric Distribution Systems and Technologies (EDST), Vienna, Austria, 8–11 September 2015; pp. 245–250. [CrossRef]
31. Hu, J.; Lan, T.; Heussen, K.; Marinelli, M.; Prostejovsky, A.; Lei, X. Robust Allocation of Reserve Policies for a Multiple-cell Based Power System. *Energies* **2018**, *11*, 381. [CrossRef]
32. Bobinaite, V.; Obushevs, A.; Oleinikova, I.; Morch, A. Economically Efficient Design of market for System Services under the Web-of-cells Architecture. *Energies* **2018**, *11*, 729. [CrossRef]
33. Wang, Q.; Zhang, C.; Ding, Y.; Xydis, G.; Wang, J.; Østergaard, J. Review of real-time electricity markets for integrating Distributed Energy Resources and Demand Response. *Appl. Energy* **2015**, *138*, 695–706. [CrossRef]
34. Lüth, A.; Zepter, M.J.; del Granado, P.C.; Egging, R. Local electricity market designs for peer-to-peer trading: The role of battery flexibility. *Appl. Energy* **2018**, *229*, 1233–1243. [CrossRef]
35. Moslehi, K.; Kumar, R. A Reliability Perspective of the smart grid. *IEEE Trans. Smart Grid.* **2010**, *1*, 57–64. [CrossRef]
36. Grijalva, S.; Tariq, M.U. Prosumer-based smart grid architecture enables a flat, sustainable electricity industry. In Proceedings of the Innovative Smart Grid Technologies Conference (ISGT), Anaheim, CA, USA, 17–19 January 2011.
37. Taft, J.; Martini, P.D.; Geiger, R. *Ultra Large-Scale Power System Control and Coordination Architecture, A Strategic Framework for Integrating Advanced Grid Functionality*; U.S. Department of Energy: Washington, DC, USA, 2014.
38. Ilo, A. Link—The smart grid Paradigm for a Secure Decentralized Operation Architecture. *Electr. Power Syst. Res.* **2016**, *131*, 116–125. [CrossRef]
39. Ilo, A. Demand response process in context of the unified *LINK*-based architecture. In *Bessède, Jean-Luc. Eco-Design in Electrical Engineering Eco-Friendly Methodologies, Solutions and Example for Application to Electrical Engineering*; Springer: Berlin/Heidelberg, Germany, 2017; ISBN 978-3-319-58171-2.

40. Mandelbrot, B. *The Fractal Geometry of Nature*; Freeman and Company: Dallas, TX, USA, 1983; ISBN 0716711869.
41. Le, T.; Retière, N. Exploring the Scale-Invariant Structure of smart grids. *IEEE Syst. J.* **2017**, *11*, 1612–1621. [CrossRef]
42. Thorp, J.S.; Naqavi, S.A. Load flow fractals. In Proceedings of the 28th IEEE Conference on Decision and Control, Tampa, FL, USA, 13–15 December 1989; Volume 2, pp. 1822–1827. [CrossRef]
43. Thorp, J.S.; Naqavi, S.A.; Chiang, N.D. More load flow fractals. In Proceedings of the 29th IEEE Conference on Decision and Control, Honolulu, HI, USA, 5–7 December 1990; Volume 6, pp. 3028–3030. [CrossRef]
44. Degaudenzi, M.E.; Arizmendi, M. Wavelet Based fractal Analysis of Electrical Power Demand. *Fractals* **2000**, *8*, 239–245. [CrossRef]
45. Huang, S.J.; Lin, J.M. Application of box counting method based fractal geometry technique for disturbance detection in power systems. In Proceedings of the IEEE Power Engineering Society General Meeting, Toronto, ON, Canada, 13–17 July 2003; Volume 3, pp. 1604–1608. [CrossRef]
46. Mamishev, A.V.; Russell, C.D.; Benner, L. Analysis of High Impedance Faults Using fractal Techniques. *IEEE Trans. Power Syst.* **1996**, *11*, 435–440. [CrossRef]
47. Ma, J.; Wang, Z. Application of Grille fractal in Identification of Current Transformer Saturation. *Power Syst. Technol.* **2007**, *31*, 84–88.
48. Wang, Z.; Ma, J.A. Novel Method to Identify Inrush Current Based on Grille fractal. *Power Syst. Technol.* **2007**, *31*, 63–68.
49. Wang, F.; Li, L.; Li, C.; Wu, Q.; Cao, Y.; Zhou, B.; Fang, B. Fractal Characteristics Analysis of Blackouts in Interconnected Power grid. *IEEE Power Eng. Lett.* **2018**, *33*, 1085–1086. [CrossRef]
50. Hou, H.; Tang, A.; Fang, H.; Yang, X.; Dong, Z. Electric power network fractal and its relationship with power system fault. *Tehn. Vjesn.* **2015**, *22*, 623–628. [CrossRef]
51. Viswanadh, S.R.; Raju, V.B. Application of fractals to power system networks. *IJSETR* **2016**, *5*, 2547–2553.
52. Sanduleac, M.; Pop, R.; Borza, P. Smart grid. Patterns of fractality. *Energetica* **2012**, *60*, 188–191.
53. Ortjohann, E.; Wirasanti, P.; Schmelter, A.; Saffour, H.; Hoppe, M.; Morton, D. Cluster fractal model—A flexible network model for future power systems. In Proceedings of the International Conference on Clean Electrical Power, Alghero, Italy, 11–13 June 2013; pp. 293–297. [CrossRef]
54. Florea, G.; Chenaru, O.; Popescu, D.; Dobrescu, R. A fractal model for Power smart grids. In Proceedings of the IEEE 20th International Conference on Control Systems and Science, Bucharest, Romania, 27–29 May 2015; pp. 572–577.
55. Miller, C.; Martin, M.; Pinney, D.; Walker, G. *Achieving a Resilient and Agile Grid*; The National Rural Electric Cooperative Association: Arlington, VA, USA, 2014; pp. 1–78.
56. COMMISSION REGULATION (EU) 2017/1485. Establishing a Guideline on Electricity Transmission System Operation. Available online: https://eur-lex.europa.eu/legal-content/EN/TXT/PDF/?uri=CELEX:32017R1485&from=EN (accessed on 10 October 2018).
57. Gupta, A. The World's Longest Power Transmission Lines. Power Technology. 2014. Available online: https://www.power-technology.com/features/featurethe-worlds-longest-power-transmission-lines-4167964 (accessed on 9 June 2016).
58. Preiss, R.F.; Warnock, V.J. Impact of Voltage Reduction on Energy and Demand. *IEEE Trans. Power Appar. Syst.* **1978**, *97*, 1665–1671. [CrossRef]
59. Schneider, K.P.; Fuller, J.C.; Tuffner, F.K.; Singh, R. *Evaluation of Conservation Voltage Reduction (CVR) on a National Level*; Pacific Northwest National Laboratory: Richland, WA, USA, 2010; pp. 1–114.
60. Bollen, M.H.J.; Sannino, A. Voltage control with inverter-based distributed generation. *IEEE Trans. Power Deliv.* **2005**, *20*, 519–520. [CrossRef]
61. Ilo, A.; Schultis, D.L.; Schirmer, C. Effectiveness of Distributed vs. Concentrated Volt/Var Local Control Strategies in Low-Voltage grids. *Appl. Sci.* **2018**, *8*, 1382. [CrossRef]
62. Ilo, A.; Schultis, D.L. Low-voltage grid behaviour in the presence of concentrated var-sinks and var-compensated customers. *Electr. Power Syst. Res.* **2019**, *171*, 54–65. [CrossRef]
63. Schultis, D.L.; Ilo, A.; Schirmer, C. Overall performance evaluation of reactive power control strategies in low voltage grids with high prosumer share. *Electr. Power Syst. Res.* **2019**, *168*, 336–349. [CrossRef]

64. Ilo, A.; Reischböck, M.; Jaklin, F.; Kirschner, W. DMS impact on the technical and economic performance of the distribution systems. In Proceedings of the CIRED 2005–18th International Conference and Exhibition on Electricity Distribution, Turin, Italy, 6 June 2005; pp. 1–5.
65. Frankhauser, P.; Tannier, C.; Vuidel, G.; Fractalyse 2.4. Research Centre ThéMA. Available online: http://www.fractalyse.org/en-home.html (accessed on 13 October 2018).
66. Bose, A. Smart Transmission grid Applications and Their Supporting Infrastructure. *IEEE Trans. Smart Grid* **2010**, *1*, 11–19. [CrossRef]
67. Marris, E. Upgrading the grid. *Nature* **2008**, *454*, 570–573. [CrossRef]
68. Amin, M. The smart grid solution. *Nature* **2013**, *499*, 145–147. [CrossRef]
69. Vaahedi, E. *Practical Power System Operation*; John Wiley & Sons: Hoboken, NJ, USA, 2014.
70. Ilo, A. The Energy Supply Chain Net. *Energy Power Eng.* **2013**, *5*, 384–390. [CrossRef]
71. Ilo, A.; Prata, R.; Iliceto, A.; Strbac, G. *Embedding of Energy Communities in the Unified LINK-Based Holistic Architecture*; CIRED: Madrid, Spain, 2019; pp. 1–5.
72. ETIP SNET. White Paper Holistic Power Systems Architectures. 2019. Available online: https://www.etip-snet.eu/white-paper-holistic-architectures-future-power-systems/ (accessed on 12 March 2019).
73. Ilo, A. Effects of the Reactive Power Injection on the grid—The Rise of the Volt/var Interaction Chain. *Smart Grid Renew. Energy* **2016**, *7*, 217–232. [CrossRef]
74. Schultis, D.L.; Ilo, A. Behaviour of Distribution grids with the Highest PV Share Using the Volt/Var Control Chain Strategy. *Energies* **2019**, *12*, 3865. [CrossRef]
75. Ilo, A.; Schaffer, W.; Rieder, T.; Dzafic, I. Dynamische Optimierung der Verteilnetze—Closed loop Betriebergebnisse. In Proceedings of the VDE Kongress, Stuttgart, Germany, 5–6 November 2012; pp. 1–6.
76. ZUQDE-project, Final Report, 1–111. 2012. Available online: https://www.energieforschung.at/assets/project/final-report/ZUQDE.pdf (accessed on 10 January 2013).

Article

Phase Synchronization Stability of Non-Homogeneous Low-Voltage Distribution Networks with Large-Scale Distributed Generations

Shi Chen [1], Hong Zhou [1], Jingang Lai [2], Yiwei Zhou [3,*], and Chang Yu [1]

1 School of Electrical Engineering and Automation, Wuhan University, Wuhan 430072, China; chenshi46@whu.edu.cn (S.C.); hzhouwuhee@whu.edu.cn (H.Z.); yuchang@whu.edu.cn (C.Y.)
2 E.ON Energy Research Center, RWTH Aachen University, 52074 Aachen, Germany; jinganglai@126.com
3 School of engineering, University of South Wales, Pontypridd CF37 1DL, UK
* Correspondence: 30014653@students.southwales.ac.uk

Received: 9 February 2020; Accepted: 3 March 2020; Published: 9 March 2020

Abstract: The ideal distributed network composed of distributed generations (DGs) has unweighted and undirected interactions which omit the impact of the power grid structure and actual demand. Apparently, the coupling relationship between DGs, which is determined by line impedance, node voltage, and droop coefficient, is generally non-homogeneous. Motivated by this, this paper investigates the phase synchronization of an islanded network with large-scale DGs in a non-homogeneous condition. Furthermore, we explicitly deduce the critical coupling strength formula for different weighting cases via the synchronization condition. On this basis, three cases of Gaussian distribution, power-law distribution, and frequency-weighted distribution are analyzed. A synthetical analysis is also presented, which helps to identify the order parameter. Finally, this paper employs the numerical simulation methods to test the effectiveness of the critical coupling strength formula and the superiority over the power-law distribution.

Keywords: large-scale; distributed generations; low-voltage active distribution network; islanded mode; non-homogeneous model; synchronization; stability

1. Introduction

Renewable energy sources have a broad prospect in power systems these days—more and more distributed generations (DGs) are integrated into the modern electricity grid. The distributed generation has the advantages of being environmentally-friendly and sustainable. In addition, two weaknesses include low output power per unit and energy output in remote areas that can not be ignored. For one thing, the low output power per unit of solar power and wind power results in a low capacity for one DG unit. This is why the distribution network requires a large number of installed DGs to provide effective output power. For another thing, the remote distribution leads to power systems usually working in the islanded operation. Different from the normal generators, there is no rotor in the DGs such as batteries and photovoltaic units. That is to say, the frequency stability is easy to be broken in the islanded microgrid composed of DG units without the rotating mass [1–3]. Hence, we will address the issue of phase and frequency stability in the islanded system.

Previous studies have based their criteria for selection on the local control of a few DGs in the ideal model [4–6]. Reference [7] shows the change of various parameters of generators in a large scale photovoltaic system and presents two different operations to research the voltage stability in an IEEE 30-bus model with a Large-Scale photovoltaic module. Indeed, the large-scale photovoltaic module can be considered as one high-capacity DG in the case of IEEE 30-bus, and no method of improving stability was given in [7]. Conventionally, when the number of composed DG increases, the active low-voltage islanded network will face the issue of high-order, multi-dimensional, and strong coupling. In brief, with a sharp increase in the number of DGs, the system will encounter the problem named "Curse of Dimensionality". Therefore, it is difficult for previous solutions that consider a few DGs to address relevant problems in current research effectively.

We note that recently there has been some advanced complex network theory that provides the stability analysis of large-scale network systems. It has wide applications in cyber science [8], medical science [9], social science [10], and so on. Nevertheless, there are a few related studies on the frequency stability of an active distribution network composed of DG units.

In addition, previous research mainly assumes that the networks are unweighted and undirected; this means that the coupling relationship between nodes has no direction, and the strength of all couplings is consistent. However, this assumption mismatches the characteristics of the real network of many cases. For instance, in an active distribution network, the coupling relationship between DGs which takes droop control is determined by the node voltage, line impedance, and droop coefficient, where the line impedance between DGs is generally determined by the distance between DGs. Apparently, the assumption that all DGs in the active distribution network is equidistant is undoubtedly harsh and impractical. On the contrary, the assumption that the coupling relationship between DGs is non-homogeneous is more realistic. Fortunately, complex network theory has made a difference in weighted networks. Reference [11] investigates the system of Kuramoto oscillators whose weight is related with a degree in the asymmetrical situation and comes to one conclusion that the critical coupling strength has a positive correlation with the degree exponent. Reference [12] indicates that the degree sequence contributes to the ability to synchronize in the mean field. Reference [13] shows that the degree distribution of the fixed system has a negative degree correlation between its nodes, which improves its synchronizability. Besides the weighted methods with the degree, some other papers pay attention to the weighted methods with frequency. Reference [14] introduces a weighting method which cares about the relationship between frequency and link to achieve the phase synchronization. Linear and nonlinear weighting procedures are applied in mean and heterogeneous systems to verify the method. Frequencies of symmetrical oscillators are studied in the bipolar model built by [15], while the relationship between frequency and coupling strength is analyzed. Reference [16] comes to the conclusion that asymmetric weights that depend on loads make contributions to the stability of the whole system. Overall, weighted methods that are based on the inherent qualities of the inverter nodes [17,18], i.e., node degrees [19], node loads [20], and network's global properties in the connection links have been discussed widely to enhance the network synchronization in recent studies.

However, the general impact of weighted coupling on network synchronization is not studied comprehensively, especially the weighted method. These weighted procedures above, unfortunately, are designed only concerned with the inherent attributes of the network, i.e., the size of the nodes degree or the edges between vertices, which limits their applications. Thus, these weighted procedures may have a significant difference from real systems. Therefore, it is important to describe such real weighted distribution methods in weighted networks.

Thereby, it is of great importance to study different weighted distribution methods of a distribution network, which has not been discussed yet. This paper employs some weighted distribution methods that are found in many realistic networks such as the link weight and investigates the phase synchronization performance in a large-scale islanded network to pave the way for the solution of the similar weighted distribution issues in the future. The innovations of this paper are as follows.

(1) We derive non-homogeneous models for simulating different weighted distribution methods in a large-scale islanded distribution network.
(2) We analytically study the effects of weighted distribution methods on the phase synchronization stability by means of explicit synchronization conditions. We illustrate our insightful results and the utility of our approach through the critical coupling strength formula.

(3) Different from other studies focusing on the weighted distribution methods which only associate with the inherent properties of the network, this paper adopts several realistic distribution styles as weighted distribution methods to research the synchronization stability in the large-scale distribution network.

The rest of this paper is organized as follows. In Section 2, we introduce correlation complex network theory. Section 3 focuses on the modeling of weighted distribution and derivation of the critical coupling strength formula. The numerical results and simulation analyses in Section 4 are used to compare the network stability of different weighting methods. Section 5 concludes this paper.

2. Correlation Theory

The structure of an active distribution network can be achieved by mirroring algebraic graph theory. Furthermore, the synchronization condition of an active distribution network will be analyzed by order parameters that are introduced thoroughly in the following. Then, the critical coupling strength of network stability can be obtained with the help of synchronization condition.

2.1. Graph Theory

The microgrid can be modeled as a graph G, where $V = \{\theta_1, \theta_2, \cdots \theta_N\}$ is the set of edges while $E \in V \times V$ is the set of edges. Generally, the connected graph can be called an undirected graph when (θ_i, θ_j) and (θ_j, θ_i) represent the same edge. The value of weight a_{ij} defines whether the graph is weighted ($a_{ij} \neq 1$) or not ($a_{ij} = 1$) and hence we have the associated adjacency matrix $A = [a_{ij}]$. Moreover, the Laplace matrix L is a symmetric zero-sum matrix of undirected graphs, where $l_{ii} = \sum_{i \neq j} a_{ij}$, $l_{ij} = -a_{ij}$. The matrix L satisfies:

$$\begin{cases} l_{ij} \leq 0, i \neq j \\ \sum\limits_{j=1}^{N} l_{ij} = 0, i = 1, \cdots, N \end{cases} \tag{1}$$

Whether the graph G is a directed graph or an undirected graph, its associated matrix is based on a directed label, which means edge $e = \{i, j\} \in E$ is represented by an ordered pair (i, j). In the undirected graph, we assign an arbitrary direction to it. For the ordered pair (i, j), we define $b_{je} = 1$ if j is the sink node of edge e, $b_{ie} = -1$ if i is the source node of edge e, and $b_{ke} = 0$ otherwise in the incidence matrix $B = [b_{ie}] \in R^{n \times m}$. In addition, the edge (i, j) satisfy $(B^T X)_e = x_j - x_i$.

2.2. Synchronization Conditions

Specifically, the phase and frequency of all DG units tend to stabilize at a fixed value when $t \to \infty$. Therefore, we focus on the synchronization process of phase and frequency to research the system stability intuitively.

If the coupling strength K between the DG nodes in a complex network is larger than the critical value K_{cr}, the DG nodes will divide into two groups, the phase of one cluster is fixed at a stable specific frequency while the other oscillators have a drift.

The order parameter is a visual signal to show the change of phase, whose definition is:

$$re^{i\psi} = \frac{1}{N} \sum_{j=1}^{N} e^{i\theta_j} \tag{2}$$

The size of r shows the percentage of stable nodes. The value is:

$$|r| = \frac{1}{N} \left| \sum_{j=1}^{N} e^{i\theta_j} \right| \tag{3}$$

where $r = 1$ means that a system enters a stable state absolutely while $r = 0$ means that a system becomes unstable.

Criterion 1 [21]: For a complex network graph G with an incidence matrix B, each pair of interconnected oscillators $\{i,j\} \in \varepsilon$ could enter frequency synchronization and phase convergence $|\theta_i - \theta_j| \leq \gamma < \pi/2$ when satisfying

$$\left\|L^{\dagger}\omega\right\|_{\varepsilon,\infty} := \left\|B^T L^{\dagger}\omega\right\|_{\infty} \leq \sin(\gamma) \tag{4}$$

where $L^{\dagger}$ is the inverse of Laplace matrix L, $\omega = (\omega_1, \cdots \omega_n)^T$ is the vector of the natural frequency of each oscillator in the network and $\|\cdot\|_{\varepsilon,\infty}$ evaluates the worst-case dissimilarity of this weighted projection.

This paper will use this synchronization condition to obtain the critical coupling strength of network synchronization under weighted conditions.

2.3. Critical Coupling Strength

In the previous subsection, the synchronization conditions recently mentioned in [22] have been introduced. In different occasions, these conditions have been proven to be applicable to oscillator models. This paper takes the Kuramoto oscillator to express the nonlinear ordinary differential equations:

$$\dot{\theta}_i = \omega_i + \sum_{j \in N_i} a_{ij} \sin(\theta_j - \theta_i), i = 1,2,3...N \tag{5}$$

The model can be expressed as a matrix:

$$\dot{\theta}_i = \omega - BW \sin(B^T\theta_i), i = 1,2,3...N \tag{6}$$

$W := diag(w)$ is the diagonal matrix of weights. In a complex network, which determines the stability of the system is not the absolute value of the phase but the phase difference. The network entry synchronizes only when two phases of coupled nodes φ_1 and φ_2 are locked at a certain ratio, which means $|n\varphi_1 - m\varphi_2| <$constant. Therefore, the phase difference vector is $\varphi(t) := B^T\theta$, and the Equation (6) can be expressed by the phase difference:

$$\dot{\varphi} = B^T\omega - B^T BW \sin(\varphi) \tag{7}$$

In the above equation, the synchronous solution can be expressed as $\dot{\varphi}=0$ or $\dot{\varphi}=\varphi^*$ for a fixed point $\varphi^* \in R^n$.

In an analysis of [23], Fazlyab found that all fixed points of Equation (7) can be expressed as:

$$\sin(\varphi^*) = W^{r-1}B^T(BW^rB^T)^{\dagger}\omega + W^{-1}Fy \tag{8}$$

where $F \in R^{m\times m-n+1}$ is a matrix whose columns span the null space of B, i.e., $BF = 0, y \in R^{m-n+1}$, and r is an arbitrary real number.

When $F = 0$ and the fixed point φ^* is unique, the particular solution $W^{r-1}B^T(BW^rB^T)^{\dagger}\omega$ is independent of r. In the special case of $r = 0$, we have that

$$\sin(\varphi^*) = B^T L^{\dagger}\omega \tag{9}$$

provided that $\|B^T L^{\dagger}\omega\|_{\infty} \leq 1$.

This section has proposed the theory about synchronization. The next section will analyze the situation of different weighting methods in star topology through algebraic graph theory, and obtain the critical coupling strength by using the synchronization condition.

3. Weighted Distribution

The weighted distribution is not only helpful to describe the features of real networks but are also crucial for controlling network synchronization. Three different weighting methods will be discussed to research the phase synchronization of a large-scale islanded network in this section. This issue takes an ideal star topology as the research object. Furthermore, a critical coupling strength formula for the weighted star network would be proposed first in this paper.

3.1. Star Topology

The purpose of this investigation is to explore the synchronization stability of large-scale grid-tied DG in islanded operation. Considering that most of the distribution networks are star topology [24,25], this paper chooses star topology as the object, as shown in Figure 1. The star network is shaped by one primary node and many branch nodes connected with it. It is characterized by high reliability, easy operation, and easy maintenance when the branch lines fail. However, this structure will consume a lot of metal and switchgear. Therefore, it is suitable for important construction sites with large capacity equipment or special places with concentrated load and wet, corrosive environments [21,26,27].

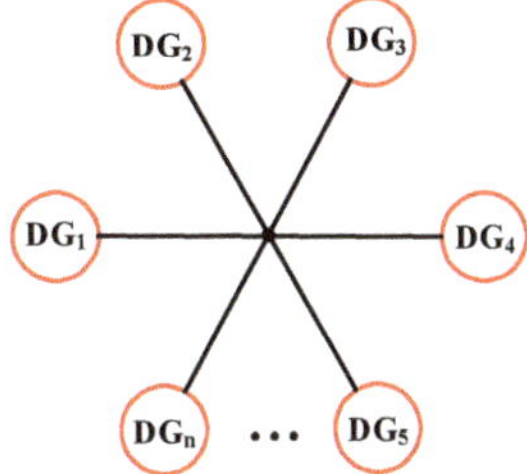

Figure 1. Star topology.

Due to the practical application of the power grid, the traditional research object of multi-machine (three to five machines) can no longer meet the actual needs. This is why this paper delivers an explicit analysis of the network stability with large-scale grid-tied DG. The number of DG N in a Large-scale model generally refers to $N \geq 200$, where this paper takes N as 201. There are no conventional generators with rotors used in this model, and all DGs are photovoltaic units. The photovoltaic community electrical diagram with star topology is depicted in Figure 2. We assume that the value of the line impedance between load nodes and the master node is consistent, and the inverter of each load node has the same impedance, so the impedance value between load node and the master node is equal. Therefore, $Z = Z_{line} + Z_{inv}$ and the admittance value is $Y = \frac{1}{Z} = \frac{1}{Z_{line}+Z_{inv}}$.

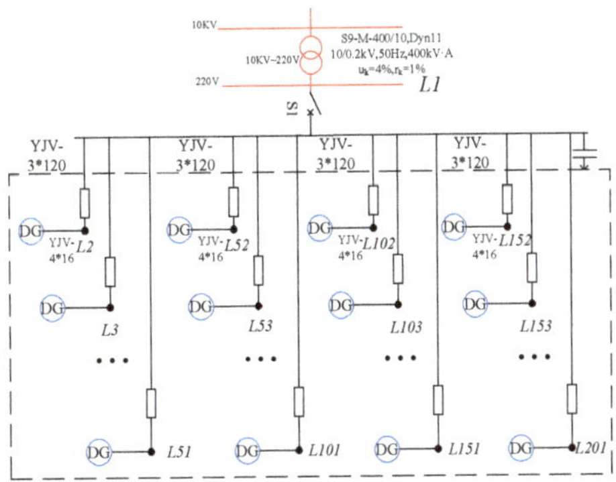

Figure 2. Star topology electrical diagram.

In Figure 2, each load node is only connected to the master node $L1$, and the impedance value between nodes $Z_{ij} = Z + Z_{inv}$ are equal. Therefore, the admittance matrix with star topology is

$$\begin{aligned}\left|Y_{ij}\right| &= \begin{bmatrix} 0 & |Y_{12}| & \cdots & |Y_{1N}| \\ |Y_{21}| & 0 & \cdots & \vdots \\ \vdots & \vdots & \vdots & \vdots \\ |Y_{N1}| & \cdots & \cdots & 0 \end{bmatrix}_{(N\times N)} \\ &= \begin{bmatrix} 0 & \left|\frac{1}{Z+Z_{inv}}\right| & \cdots & \left|\frac{1}{Z+Z_{inv}}\right| \\ \left|\frac{1}{Z+Z_{inv}}\right| & 0 & \cdots & \cdots \\ \vdots & \vdots & \vdots & \vdots \\ \left|\frac{1}{Z+Z_{inv}}\right| & \cdots & \cdots & 0 \end{bmatrix}_{(N\times N)}\end{aligned} \tag{10}$$

3.2. The General Formula for Weighted Distribution

The DG units with droop control can be described as a Kuramoto oscillator [28–30]:

$$\begin{aligned}\frac{d\theta_i}{dt} = \dot{\theta}_t = \omega_i - \omega_\theta = -m_i\left(p_i - p_{i\theta}\right) = m_i p_{i\theta} - m_i p_i \\ = m_i p_{i\theta} - m_i E_i \textstyle\sum_{j=1}^{n} \left|Y_{ij}\right| E_j \sin\left(\theta_i - \theta_j\right)\end{aligned} \tag{11}$$

where $m_i p_{i\theta}{=}\omega_i$, ω_i is the rated frequency of inverter in the ith DG node. In addition, the value of ω_i is chosen from the probability density function $g(\omega)$. When the standard frequency is 50 Hz, we take $g(\omega)$ as the symmetric normal distribution of 50, therefore $g(\omega){\sim}\ (50,1)$.

In the above equation, the weight between DG_i and DG_j nodes is obtained by the droop coefficient d_i, the voltage of node E_i, E_j, and the admittance $\left|Y_{ij}\right|$. At this point, the coupling strength

$$K_i = d_i \left|Y_{ij}\right| E_i E_j = k m_i$$

For the sake of simplicity, we divide the weight between the DG nodes into two parts. One is the coupling strength k, and the other is the weight index m_i that conforms to some kind of weighted distribution. Therefore, the weight of the directed edge $\{i,j\}$ is determined by the weight index m_i while the weight of the edge $\{j,i\}$ is determined by the weight index m_j, so the weight distribution of one network can be decided by the distribution of weight index m_i. In addition, we can synchronously adjust the size of all m_i values to change the size of the weight. Thus, we have the weight diagonal matrix $W = k \times diag(m)$, where m_1 is the weight when a load node is the sink node and the master node is the source node, $m_2 \cdots m_n$ is the weight when the master node is the sink node and a load node is the source node. Thus, the weight distribution of the network can be simulated by adjusting the value of the weight index:

$$
\begin{aligned}
A &= \begin{bmatrix} 0 & km_1 & \cdots & km_1 \\ km_2 & 0 & \cdots & 0 \\ \vdots & \vdots & \vdots & \vdots \\ km_n & 0 & \cdots & 0 \end{bmatrix}_{(201\times201)} \\
D &= \begin{bmatrix} (n-1)\times km_1 & 0 & \cdots & 0 \\ 0 & km_2 & \cdots & 0 \\ \vdots & \vdots & \vdots & \vdots \\ 0 & \cdots & \cdots & km_n \end{bmatrix}_{(201\times201)} \\
L &= \begin{bmatrix} (n-1)\times km_1 & km_1 & \cdots & km_1 \\ km_2 & -km_2 & \cdots & 0 \\ \vdots & \vdots & \vdots & \vdots \\ km_n & 0 & \cdots & -km_n \end{bmatrix}_{(201\times201)}
\end{aligned}
\tag{12}
$$

In a star topology whose N is 201, the incidence matrix is

$$
B = \begin{bmatrix} 1 & 1 & \cdots & 1 & 1 \\ -1 & 0 & 0 & \cdots & 0 \\ \vdots & \vdots & \vdots & \vdots & \vdots \\ 0 & 0 & \cdots & -1 & 0 \\ 0 & 0 & \cdots & 0 & -1 \end{bmatrix}_{(201\times200)}
\tag{13}
$$

3.3. *Weighted Distribution Model*

Different weighting methods have different performances in different topology, but the topology of an actual power grid is fixed and difficult to adjust. Thus, the weighting methods, which are associated with the degree distribution and node strength, are not considered in this paper. Only the weighting method associated with the distribution of coupling strength is considered in this paper.

In practice, the error occurs in the installation process often obey Gaussian distribution, the weight distribution of complex network is mostly in line with the power-law distribution, and the frequency-weighted distribution can reflect some peculiarities of the actual network. Thus, this paper discusses the weighting methods of Gaussian distribution, power-law distribution, and frequency-weighted network, and the aforementioned methods can be simulated by adjusting the weight index.

3.3.1. Gaussian Distribution

At first, this paper considers the case that the weight distribution method is the same as the rated frequency of an inverter distribution method so that the distribution function $g(m)$ of m_i is the same as the distribution function $g(\omega)$ of ω_i; both are Gaussian distribution. Unlike the rated frequency of the inverter being a symmetric normal distribution of 50, the weight distribution is a symmetric normal distribution of the y-axis. Its Kuramoto oscillator model is:

$$
\begin{aligned}
\dot{\theta}_i &= \omega_i + kg(m) \sum_{j=1}^{N} a_{ij} \sin(\theta_i - \theta_j), i = 1, \ldots, N \\
&= \omega_i + k\,|m_i| \sum_{j=1}^{N} a_{ij} \sin(\theta_i - \theta_j), i = 1, \ldots, N.
\end{aligned}
\tag{14}
$$

$a_{ij} = 1$ when DG_i has coupling relationship with DG_j, otherwise $a_{ij} = 0$.

3.3.2. Power-Law Distribution

Power-law distribution has extensive application in complex network theory. Barabasi and Albert pointed out that the degree distribution of complex networks is the power-law distribution in [31]. Then, Chunguang Li and Guanrong Chen verified that the weight distribution is a power-law distribution in a general complex network. Furthermore, according to the research of Barrat et al. in [32], the node strength distribution in the weighted network obeys the power-law distribution as well. In this paper, only the weighted distribution of weights is discussed. Therefore, it is worth mentioning that the assumed distribution function $g(m)$ of m_i satisfies the power-law distribution, which means $g(m) \sim m^{-\beta}$, and

$$\begin{aligned} \dot{\theta} &= \omega_i + k g(m) \sum_{j=1}^{N} a_{ij} \sin(\theta_i - \theta_j), i = 1, \ldots, N \\ &= \omega_i + k m_i \sum_{j=1}^{N} a_{ij} \sin(\theta_i - \theta_j), i = 1, \ldots, N \end{aligned} \tag{15}$$

3.3.3. Frequency-Weighted Network

The frequency-weighted network reflects the feature of several natural and social networks. In fact, a power grid could be expressed as a Kuramoto oscillator model, and the weight of the coupling strengths are obtained by their natural frequency. In this case, the weighted coupling coefficients are related to DGs' own inverter rated frequency rather than the topology of a network. We added the inverter rated frequency to the dynamic model as follows:

$$\dot{\theta}_i = \omega_i + \frac{k\,|\omega_i|}{\sum_{j=1}^{N} a_{ij}} \sum_{j=1}^{N} a_{ij} \sin(\theta_i - \theta_j), i = 1, \ldots, N \tag{16}$$

3.4. Critical Coupling Strength Formula

Reference [23] pointed out that any fixed point in Equation (7) has local exponential stability when $\|\varphi^*\|_\infty \leq \pi/2$. For the sake of simplicity, we decompose the matrix $A = k \times A'$, where A' is the adjacency matrix entirely determined by weight indexes, and k is the coupling strength. The in-degree matrix D is a diagonal matrix, whose values in each row are the row sum of corresponding rows of A matrix. Certainly, the D matrix can be decomposed into $D = k \times D'$, where D' is the in-degree matrix entirely determined by weight indexes. At the same time, the Laplacian matrix $L = D - A$ can be decomposed into $L = k \times L'$, where L' is the Laplacian matrix entirely determined by weight indexes. Furthermore, Equation (9) can be decomposed into:

$$\sin(\varphi^*) = k^{-1} B^T (L')^{\dagger} \omega \tag{17}$$

There is $\left\|\sin(B^T\theta^*)\right\|_\infty = 1$ when the network enters a critical stable state, and then the critical coupling values of star topology in different weighting methods can be transformed into the following formula:

$$k_c = \left\| B^T (L')^{\dagger} \omega \right\|_\infty \tag{18}$$

The calculation of the critical coupling value in different weighting methods is described in detail in Section 4.

4. Simulation

In this section, three weighted methods from above are simulated, and the proposed formula of critical coupling (18) is verified.

4.1. Case 1

It is inevitable that deviations will generate in the process of installing the grid-tied DG inverters in an active distribution network, and the distribution of deviations often conforms to the Gaussian distribution.

In this case, the assumed actual parameters are reported in Table 1:

Table 1. Related parameters in Case 1.

Type	Parameter
wire type	YJV-3
wire section	3×120 mm^2
wire resistance	0.153/km
wire length	100 km
number of DG	51/101/201
wire type in inverter	YJV-4
wire section in inverter	4×16 mm^2
wire resistance in inverter	1.15/km
wire length in inverter	10 km
rated frequency of inverter	50 Hz
distribution of inverter rated frequency	Gaussian distribution
distribution of weight index	Gaussian distribution
voltage classes	220 V

4.1.1. Weighted Undirected Network

Let us now turn to the case of the undirected situation, the weight distribution of edges in this situation conforms to the Gaussian distribution $X \sim N(0,1)$, and source point and sink point of each edge are arbitrary. When N takes 201, and the number of edges takes 200 in this case; meanwhile, the distribution is depicted in Figure 3. Furthermore, Figure 4 shows the probability density function of this Gaussian distribution, and it is obvious that the distribution law conforms to the (0,1) distribution.

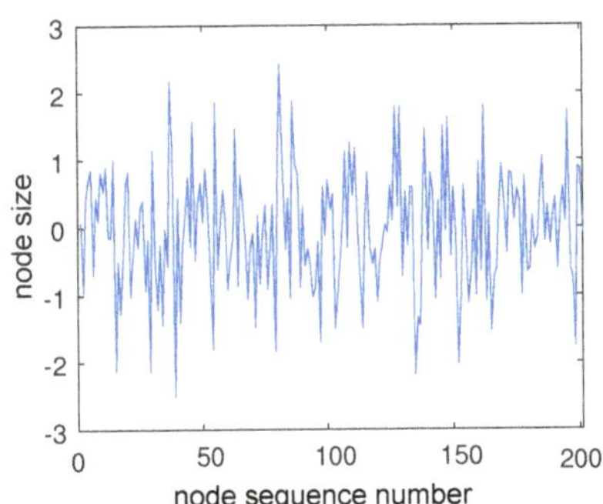

Figure 3. Nodes in Gaussian distribution.

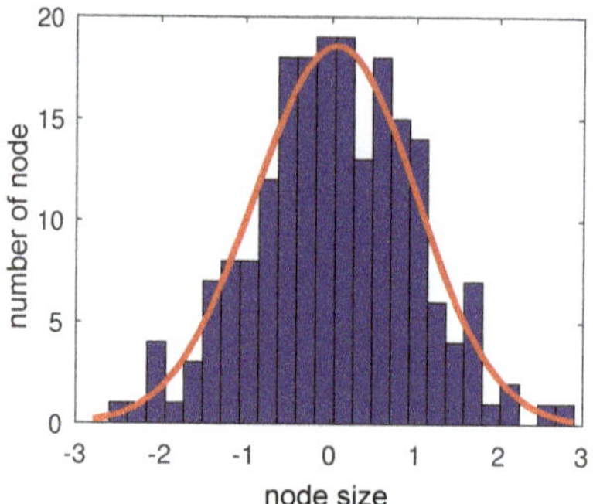

Figure 4. Probability density function of Gaussian distribution.

Thus, the ordinary differential equation of oscillator θ_i is

$$\dot{\theta} = \omega_i + kX \sum_{j=1}^{N} a_{ij} \sin(\theta_i - \theta_j), i = 1, \ldots, N \tag{19}$$

where X is the weight of the corresponding edge.

Turn now to verifying the critical coupling strength formula of the weighted situation proposed in Section 3. In this case,

$$L = \begin{bmatrix} -\sum_{1}^{200} X & X_1 & \cdots & X_{200} \\ X_1 & -X_1 & \cdots & 0 \\ \vdots & \vdots & \vdots & \vdots \\ X_{200} & 0 & \cdots & -X_{200} \end{bmatrix}_{(201 \times 201)} \tag{20}$$

Furthermore, the related parameters in this case are depicted in Table 1, and the weight $k = \left| \frac{a}{Z + Z_{\text{inv}}} \right| \times E^2 = \frac{0.1 \times 220^2}{100 \times 0.153 + 10 \times 1.15} = 180$. The phase and frequency stability are shown in Figures 5 and 6, and it can be seen that the phases of DG nodes change at the same rate and the frequency of DG nodes fixed in one constant value.

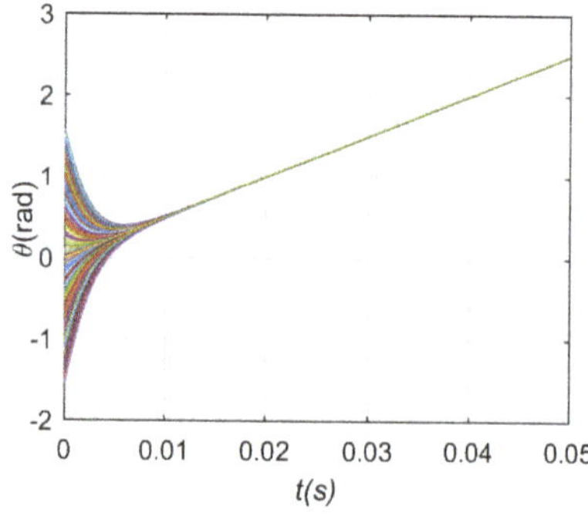

Figure 5. Phase stabilization process of undirected network in case 1.

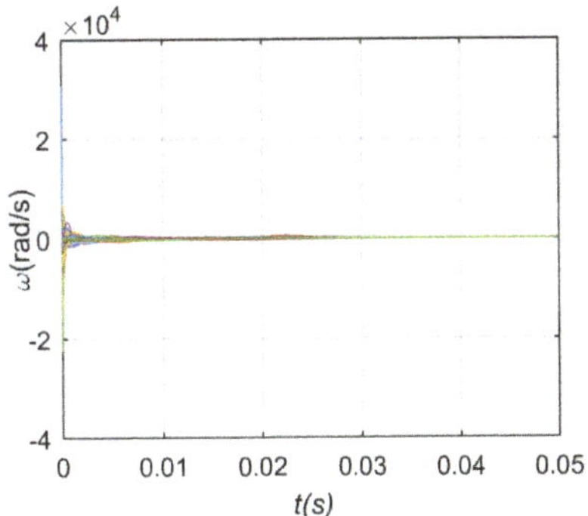

Figure 6. Frequency stabilization process of undirected network in case 1.

In this paper, order parameters are used to reflect the synchronization of DGs in the network. Figure 7 shows the relationship between the coupling strength K and the order parameter r when N has different values, which reflect the change of network synchronization with the coupling strength. It is obvious that the network stability is the best when the value of N is 51, followed by 101. When N takes 201, the network has the worst stability. It can be seen that the stability of the network decreases with the increase of N, which is not consistent with the situation in which the stability does not change with the value of N when the star topology is an undirected network.

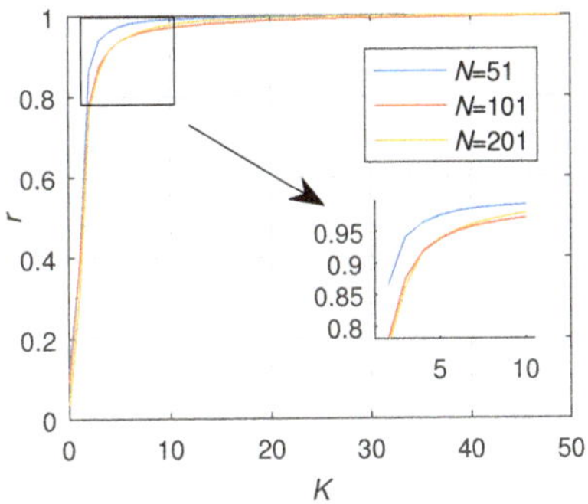

Figure 7. The stability of the weighted undirected network when N is 51,101,201.

4.1.2. Weighted Directed Network

Thus far, this paper has focused on an ideal situation that the weighted distribution of the undirected edge in the network is Gaussian distribution. Let us now turn to consider the weighted directed case. In a power grid, each DG can become not only the power transmitter but also the power receiver. Therefore, it is of great practical value to consider the different weights of edges when they take different directions between coupling DG nodes. That is, the edge weight is the weight index of the master node when the master node is the source point, while the edge weight is the weight index of the load node when the load node is the source point:

Since the inverter rated frequency of the DG unit also conforms to the Gaussian distribution, this paper assumes that the weight distribution of edges is the same type as the distribution of inverter rated frequency. Namely, the ordinary differential equation with the oscillator θ_i is

$$\dot{\theta} = \omega_i + k\,|\omega_i| \sum_{j=1}^{N} a_{ij} \sin(\theta_i - \theta_j), i = 1, \ldots, N \tag{21}$$

This paper takes the weight index of the load node as the Gaussian distribution and the weight index of the master node as the mean value of the weight index of each load node to realize the power balance. Thus, there is

$$\begin{aligned} L &= \begin{bmatrix} (n-1)*m_1 & m_1 & \cdots & m_1 \\ m_2 & -m_2 & \cdots & 0 \\ \vdots & \vdots & \vdots & \vdots \\ m_n & 0 & \cdots & -m_n \end{bmatrix}_{(201\times 201)} \\ &= \begin{bmatrix} -\sum_2^{201} m_i & \frac{\sum_2^{201} m_i}{200} & \cdots & \frac{\sum_2^{201} m_i}{200} \\ m_2 & -m_2 & \cdots & \cdots \\ \vdots & \vdots & \vdots & \vdots \\ m_n & 0 & \cdots & -m_n \end{bmatrix}_{(201\times 201)} \end{aligned} \tag{22}$$

Frequency deviation will lead to power imbalance and even instability of the whole system. The phase and frequency stabilization processes in the directed situation are shown in Figures 8 and 9 when $k = 180$. Obviously, the coupling strength k in this case is much greater than the critical coupling strength required for network stability. Let us now turn to the critical coupling strength in the case of Gaussian distribution. When $N = 201$, the value of L matrix can be substituted for formula (18) to obtain the critical coupling strength $K_C = 6.7$; when N is 51 and 101, the critical coupling strength $K_C = 3.7$ and 4.46 can be obtained from the calculation formula, respectively. Figure 10 shows that, when N takes 201, the network enters synchronization rapidly before $K < 6.7$, while the network is nearly stable after $K > 6.7$. This result is basically consistent with the result obtained by formula (4-1). Similarly, when the value of N is 51 and 101, the network becomes stable after $K_C = 3.7$ and 4.46, respectively. Thus far, two conclusions have been shown. The first is that the validity of the critical coupling strength formula (18) in the directed weighted network is verified, and the second is that the network stability decreases with the increasing number of nodes in the island star topology directed weighted network.

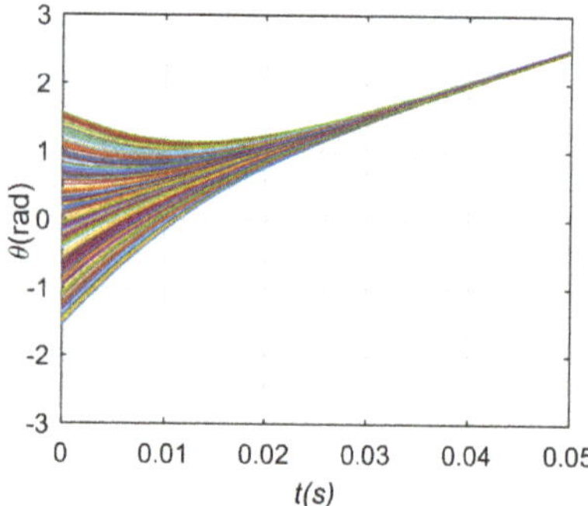

Figure 8. Phase stabilization process of directed network in case 1.

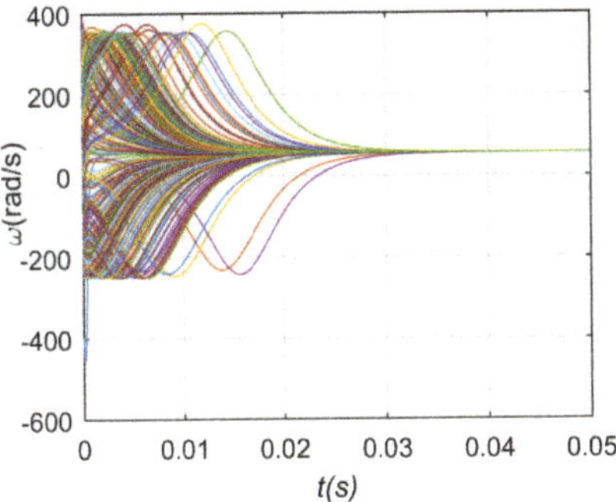

Figure 9. Frequency stabilization process of directed network in case 1.

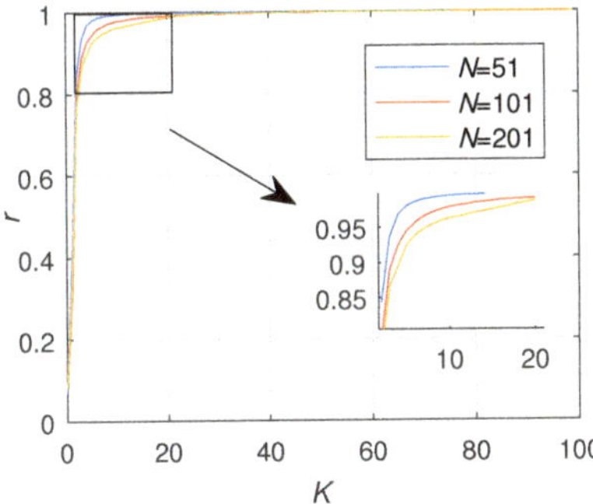

Figure 10. The stability of the weighted directed network when N is 51,101,201.

4.2. Case 2

In particular, previous studies show that most of the nodes have a degree that is a less than average value while a few nodes have large connections in the real network. The study of Chunguang Li and Guanrong Chen [17] confirmed that the distribution of connection weights would follow the power-law distribution in general actual complex networks. Thus, it is important to research the stability of a weighting case that conforms to the power-law distribution.

In this case, the assumed actual parameters are reported in Table 2:

Table 2. Related parameters in Case 2.

Type	Parameter
wire type	YJV-3
wire section	3×120 mm^2
wire resistance	0.153/km
wire length	100 km
number of DG	51/101/201
wire type in inverter	YJV-4
wire section in inverter	4×16 mm^2
wire resistance in inverter	1.15/km
wire length in inverter	10 km
rated frequency of inverter	50 Hz
distribution of inverter rated frequency	Gaussian distribution
distribution of weight index	Power-law distribution
voltage classes	220 V

4.2.1. Weighted Undirected Network

In this case, we assume that the weight distribution of 200 edges conforms to the power-law distribution, as shown in Figure 11. Then, Figure 12 shows the probability density function diagram. It can be seen that, with the increase of abscissa, the probability of weight distribution is lower. Furthermore, the probability of randomly selecting a node with greater weight is negatively correlated with the number of the same weight. The power-law index in Figure 11 is 2. The distribution of nodes depicted as a line whose slope is negative in the double logarithmic diagram, as demonstrated in Figure 12. This distribution has an asymptote of the black dotted line, where the exponent determines the rate of decay.

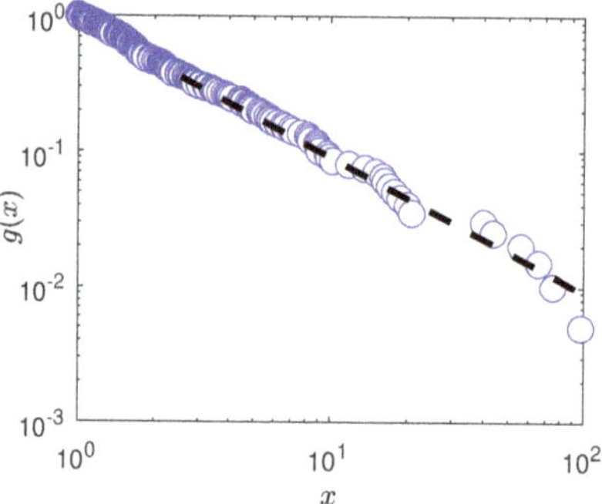

Figure 11. Nodes in power-law distribution.

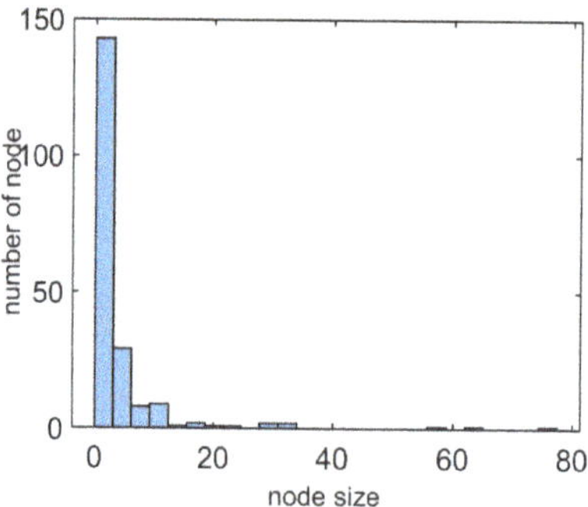

Figure 12. Probability density function of power-law distribution.

The weighted distribution in this case follows $g(X) \sim X^{-\beta}$, which means that the ordinary differential equation of oscillator θ_i is

$$\dot{\theta} = \omega_i + kX \sum_{j=1}^{N} a_{ij} \sin(\theta_i - \theta_j), i = 1, \ldots, N \tag{23}$$

where X is the weight of the corresponding edge.

Like the weighted undirected network in case 1, the weighted distribution in this case is reflected by the value of the edge weight, so the Laplace matrix is

$$L = \begin{bmatrix} -\sum_{1}^{200} X & X_1 & \cdots & X_{200} \\ X_1 & -X_1 & \cdots & 0 \\ \vdots & \vdots & \vdots & \vdots \\ X_{200} & 0 & \cdots & -X_{200} \end{bmatrix}_{(201\times 201)} \quad (24)$$

The same as the coupling strength value of case 1, the phase and frequency stabilization process diagram are shown in Figures 13 and 14. In addition, the influence of different power exponents on network stability is shown in Figure 15. Figure 15 shows the network synchronization process when power-law index β grades –5, –3, –2, 2, 3, 5. It is obvious that the critical coupling strength of the system when the power-law index is positive does not differ greatly, while the positive power-law index values have significantly better stability than the negative values. This paper comes to a conclusion that the positive and negative value of power-law index rather than the size of the Power-law index has a key influence on the stability of the network when ignoring the deviations generated in the calculation process.

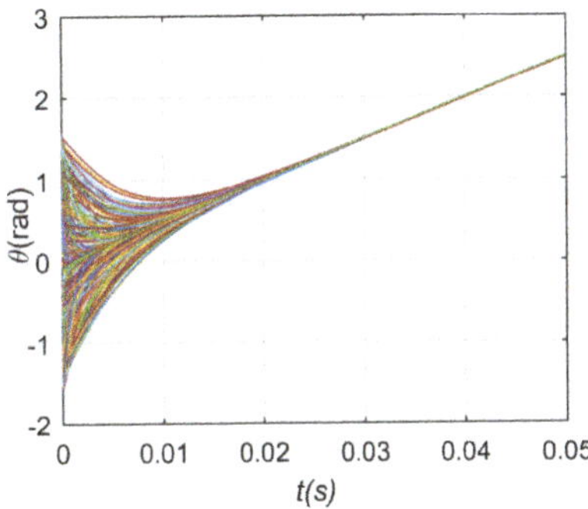

Figure 13. Phase stabilization process of undirected network in case 2.

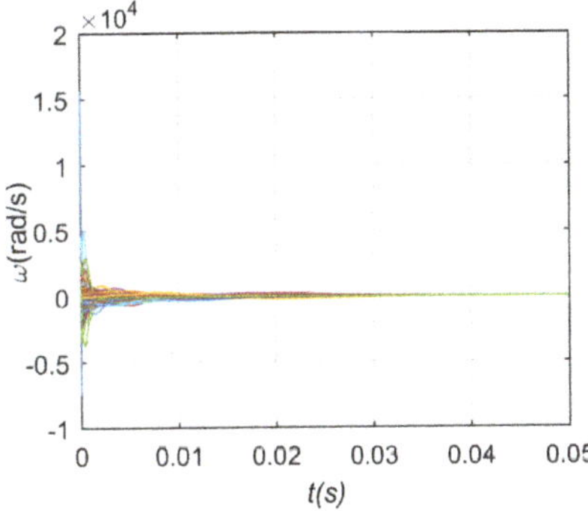

Figure 14. Frequency stabilization process of undirected network in case 2.

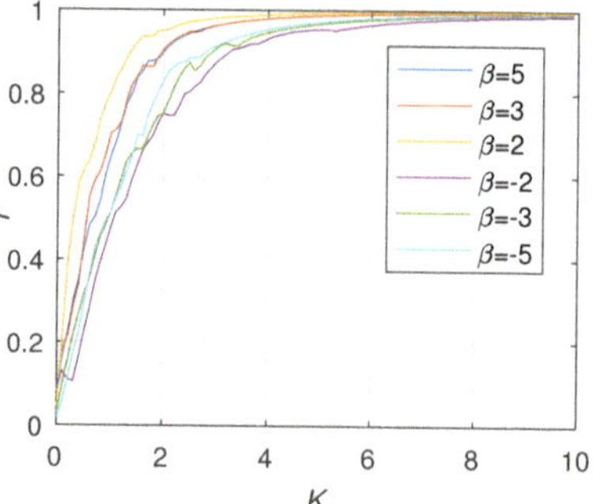

Figure 15. The stability of the weighted undirected network when β is $\pm 2, \pm 3, \pm 5$.

The simulation analysis of the different number of nodes N is shown in Figure 16. When the index is 2, the value of N is negatively correlated with network stability. The conclusion also reflects the analysis of the relationship between the node number and the network stability in the previous case.

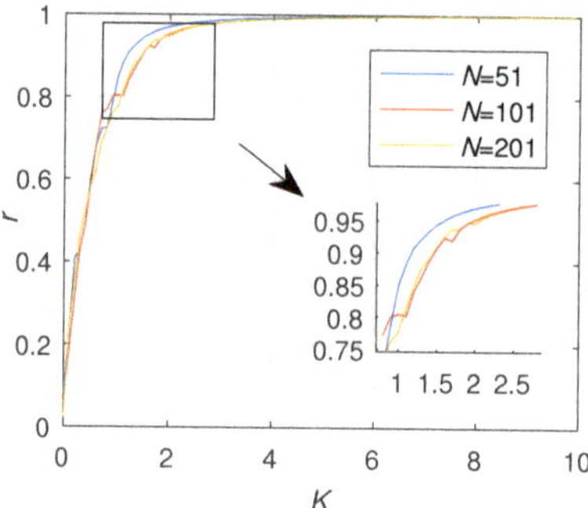

Figure 16. The stability of the weighted undirected network when N is 51,101,201($\beta = 2$).

4.2.2. Weighted Directed Network

Turn now to the directed weighted case. Similar to case 1, the method of adjusting the weight index is used to complete the weight distribution conforms to the power-law distribution $g(m) \sim m^{-\beta}$, and the ordinary differential equation is given in the formula (25):

$$\dot{\theta} = \omega_i + kg(m) \sum_{j=1}^{N} A_{ij} \sin(\theta_i - \theta_j), i = 1, \ldots, N \tag{25}$$

Similar to the weighted directed network in case 1, the weight index of the master node in this case is the mean value of the weight index of each load node considering the problem of power balance. Therefore, the Laplace matrix in the network with 201 nodes is:

$$L = \begin{bmatrix} (n-1)*m_1 & m_1 & \cdots & m_1 \\ m_2 & -m_2 & \cdots & 0 \\ \vdots & \vdots & \vdots & \vdots \\ m_n & 0 & \cdots & -m_n \end{bmatrix}_{(201\times201)} = \begin{bmatrix} -\sum_2^{201} m_i & \frac{\sum_2^{201} m_i}{200} & \cdots & \frac{\sum_2^{201} m_i}{200} \\ m_2 & -m_2 & \cdots & \cdots \\ \vdots & \vdots & \vdots & \vdots \\ m_n & 0 & \cdots & -m_n \end{bmatrix}_{(201\times201)} \quad (26)$$

The phase and frequency stabilization process in the directed situation are shown in Figures 17 and 18 when the value of power-law index beta is 2, and coupling strength k is 180. When the value of the power-law index respectively takes –5, –3, –2, 2, 3, 5, the critical coupling strength K_C of network is respectively 1.4976, 1.4353, 1.9639, 1.4976, 1.4353, and 0.7771 by formula (18). These critical coupling strengths are nearly consistent with the critical value depicted in Figure 19. In addition, the weighted directed network has similar performance to the weighted undirected network, which means that the positive power-law index value has better network stability performance than the negative one, while the size of the power-law index value has almost no effect. Apparently, in the case of power-law distribution, the stability of the network is obviously better than that in the case of Gaussian distribution. When the Power-law index is 2, the network can be stable only on the condition that the coupling strength is greater than the critical value K_C = 0.5475.

Furthermore, the relationship between the number of DG nodes and network stability is shown in Figure 20. The negative correlation between DG number and critical coupling strength is still evident.

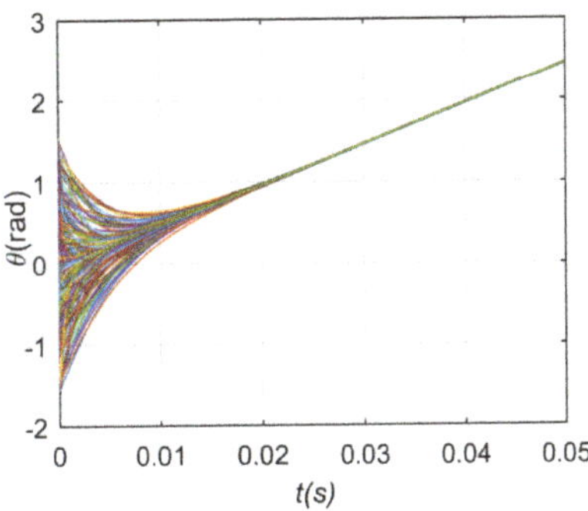

Figure 17. Phase stabilization process of directed network in case 2.

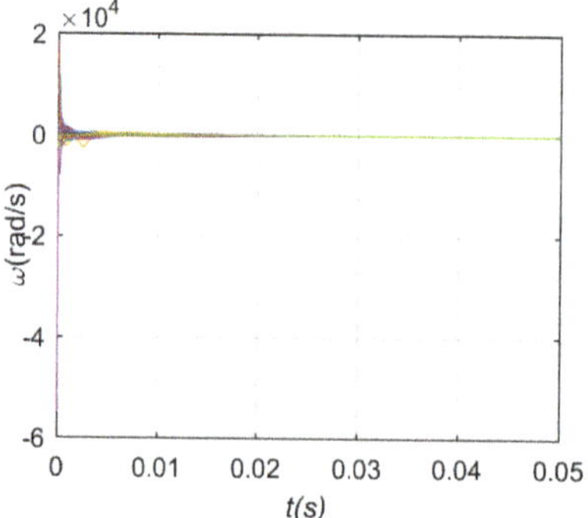

Figure 18. Frequency stabilization process of directed network in case 2.

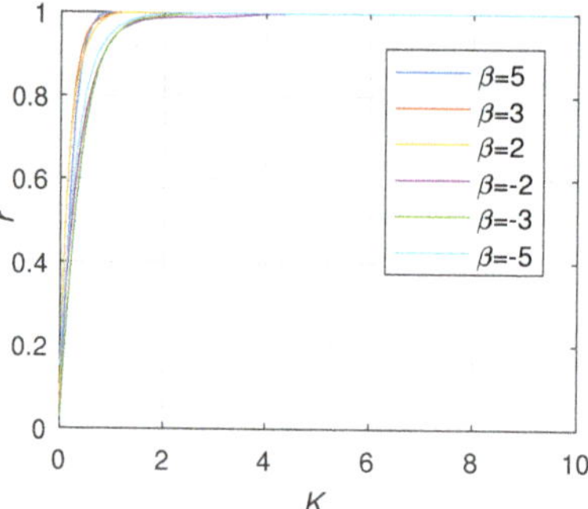

Figure 19. The stability of the weighted directed network when β is $\pm 2, \pm 3, \pm 5$).

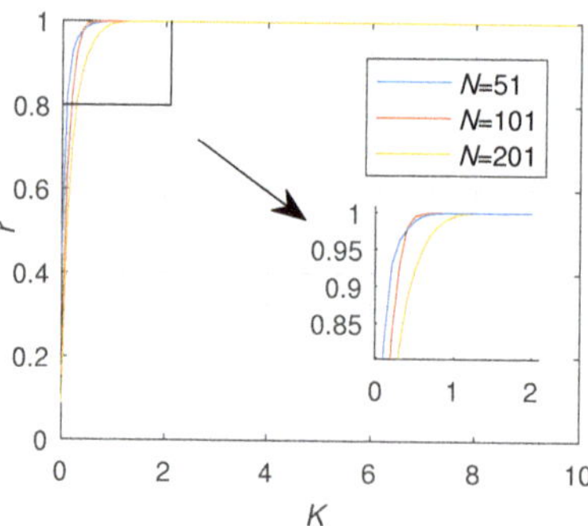

Figure 20. The stability of the weighted directed network when N is 51,101,201 (β = 2).

4.3. Case 3

The following case will discuss the Frequency-weighted situation. The natural frequency is inverter rated frequency in the Kuramoto oscillator model. Thus, case 3 chooses the absolute ratio between the inverter rated frequency of one DG node and its in-degree as the weight; there is the ordinary differential equation:

$$\dot{\theta} = \omega_i + \frac{k\,|\omega_i|}{\sum_{j=1}^{N} A_{ij}} \sum_{j=1}^{N} A_{ij} \sin(\theta_i - \theta_j), i = 1, \ldots, N \tag{27}$$

The weighted distribution in this case is different from the previous two cases, whose weight index is not a random value which obeys a certain distribution. The Laplace matrix in this case is

$$L = \begin{bmatrix} -|\omega_1| & \frac{|\omega_1|}{200} & \cdots & \frac{|\omega_1|}{200} \\ |\omega_2| & -|\omega_2| & 0 & 0 \\ \vdots & \vdots & \vdots & \vdots \\ |\omega_{n-1}| & 0 & \cdots & 0 \\ |\omega_n| & 0 & \cdots & -|\omega_n| \end{bmatrix}_{(201\times 201)} \tag{28}$$

In case 3, the phase and frequency stabilization process diagram are shown in Figures 21 and 22, which are similar to the above two cases. According to formula (18), the critical coupling strength of this case is $K_C = 12.89$ when $N = 201$, and the network synchronization progress is shown in Figure 23. It can be seen that the weighted method in this case is not as stable as the Gaussian distribution and the power-law distribution.

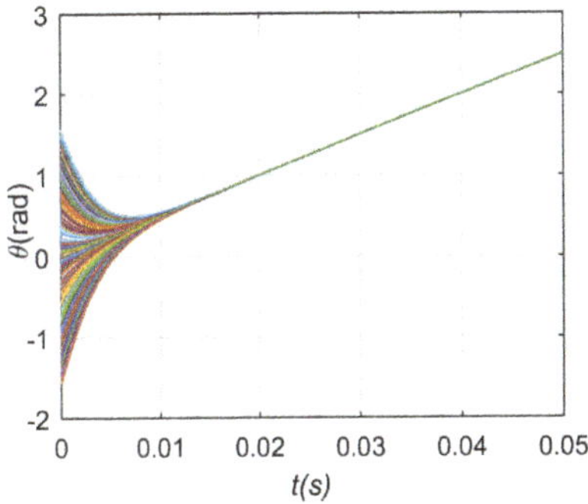

Figure 21. Phase stabilization process in case 3.

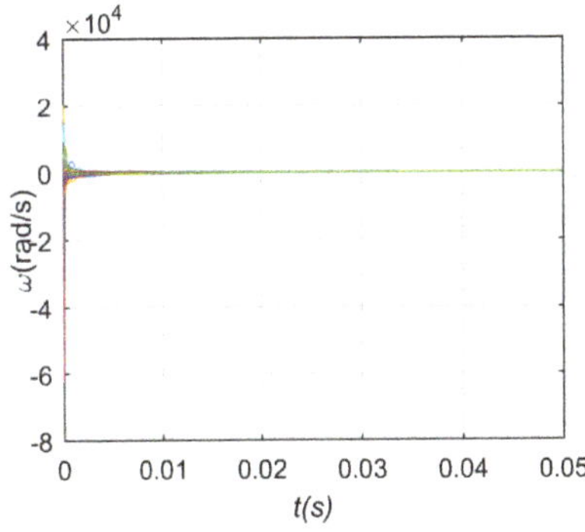

Figure 22. Frequency stabilization process in case 3.

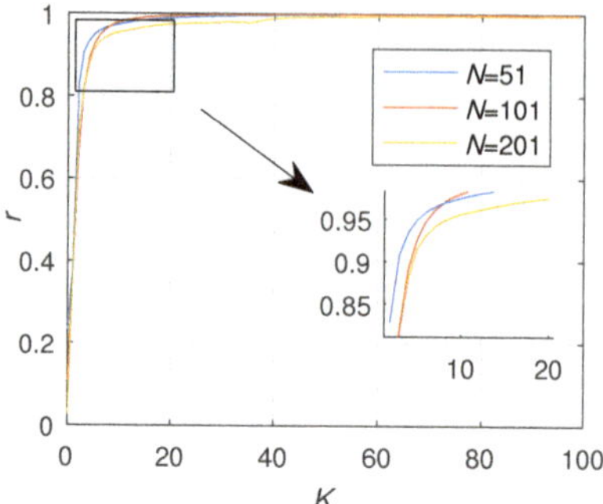

Figure 23. Synchronization progress when N is 51,101,201 in case 3.

4.4. Stability Analysis

Through comparing the critical coupling strength of three cases as illustrated in Figure 24, we find that the power-law distribution has the best stability in different weighted distribution methods. Moreover, the second largest eigenvalue λ_2 or the eigenvalue proportion λ_N/λ_2 in adjacency matrix A could describe the synchronization ability of the system. The network coherence is proportional to λ_2 and inversely proportional to λ_N/λ_2. Table 3 shows that the λ_2 of power-law distribution has the smallest value, which means it has the best synchronization ability, followed by the Gaussian distribution and frequency-weighted distribution. Thus, the results of Table 3 and Figure 24 confirm each other. In addition, the contrastive analysis of weighted directed network and weighted undirected network show that the weighted directed network which conforms to the Gaussian distribution or power-law distribution is easier to get into synchronization than the weighted undirected network. Furthermore, by studying the network with different numbers of DG, this paper finds that the stability of the weighted network decreases with an increase of N, which is different from the average field. For the weighting method of power-law distribution, the size of the power-law index β is not significant to the stability of the network, but the sign of the power-law index β is not negligible to the stability of the network.

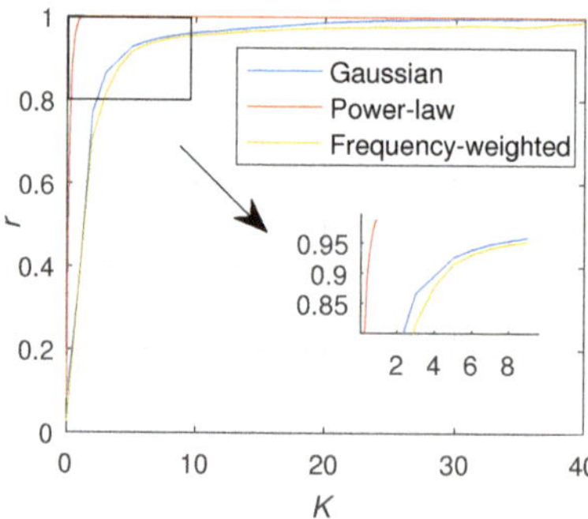

Figure 24. Synchronization progress of different weighting distributions.

Table 3. Spectrum analysis.

Distribution Mode	λ_2	λ_N	λ_N/λ_2
Gaussian distribution	–1.005	–557.16	554.38
Power-law distribution	–1.006	–436.59	433.98
Frequency-weighted	–1.0007	–686.99	686.51

4.5. Further Analysis of Weighted Network in the Actually Distributed Network

Actually, the topology structure of an actual power grid network often isn't a standard star topology. The standard star topology is an ideal assumption. In order to verify the stability analysis from the above subsection, we take the dynamic IEEE test systems [33] and the Iceland 189-node distributed grid [34] as a research object. The synchronizing processes of Gaussian distribution, power-law distribution, and frequency-weighted network are shown in Figures 25 and 26.

4.5.1. Dynamic IEEE Test Systems

Three different weighted distribution methods have been applied to the dynamic IEEE models, including 9-bus, 14-bus, 57-bus, and 118-bus modified test systems. The diagram of coupling strength and order parameter is shown in Figure 25, where the labels G, P, and F represent Gaussian distribution, Power-law distribution and frequency-weighted distribution, respectively, and the network with the Power-law distribution method always enter the stable state at first. The value of the critical coupling strength of three weighted distribution methods in the dynamic IEEE test systems is shown in Table 4, where each value is calculated by Equation (18). It shows that the network with the power-law distribution method always has the least critical coupling strengths, which corresponds to the stability performance in Figure 25.

Table 4. K_c of three weighted distribution methods in different IEEE systems.

IEEE Test Systems	**9-bus**	**14-bus**	**57-bus**	**118-bus**
Gaussian distribution	3.4	4.2	47	56.1
Power-law distribution	1.1	1.5	10.7	16.1
Frequency-weighted	4.5	4.9	78.4	113.4

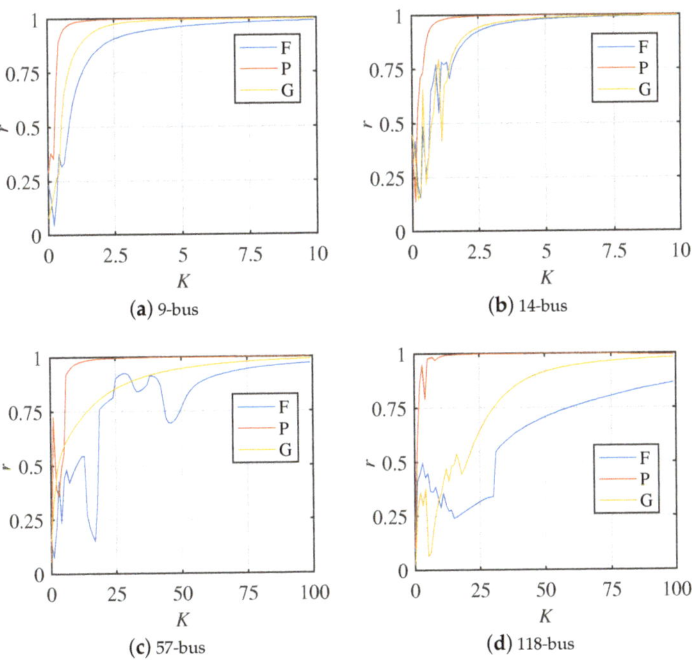

Figure 25. Synchronization progress of different weighting methods in the dynamic IEEE test systems.

4.5.2. Iceland 189-Nodes Distribution Grid

Figure 26 shows the results of the stability performance of three distribution methods in an Iceland 189-nodes distribution grid. It can be seen that the power-law distribution undoubtedly has the best stability performance. Second is the frequency-weighted network. Moreover, the Gaussian distribution needs a long process to enter synchronization. These results further support the conclusions of the above subsection.

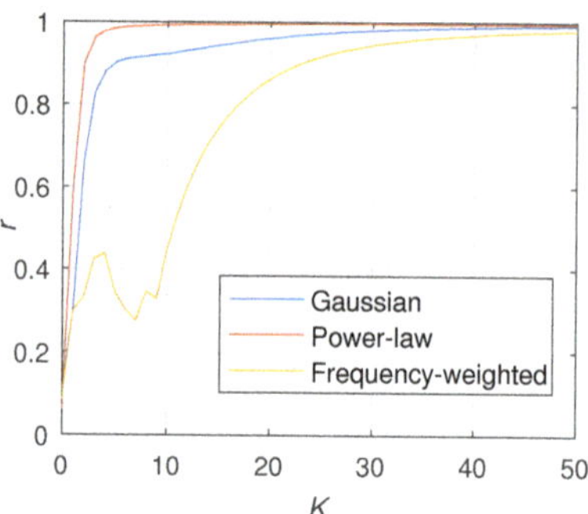

Figure 26. Synchronization progress of different weighting methods in the Iceland 189-node distributed grid.

5. Conclusions

Above all, this paper analyzes the stability of large-scale grid-tied DG in a non-homogeneous active power network by using complex networks theory.

Then, three different weighting distribution methods, including Gaussian distribution, power-law distribution, and frequency-weighted distribution, are analyzed. In addition, the critical coupling strength formula for the weighted situation $k_c = \left\| B^T(L)^{\dagger}\omega \right\|_{\infty}$ has been proposed and verified. The critical coupling strength obtained from this formula conforms to the reality when ignoring tiny deviation.

Among the three different weighting methods, the power-law distribution has the best stability. Moreover, the difference of power-law index value has little effect on the network stability, but the sign of power-law index has quite an effect on the system stability. The second is the Gaussian distribution, which is obviously less stable than the power-law distribution. Finally, the frequency-weighted network has the worst synchronization performance, which is significantly inferior to the other two. In addition, further analysis of the IEEE dynamic test systems and the Iceland 189-node distributed grid has proved the better stability of power-law distribution in this paper.

Author Contributions: Conceptualization, S.C. and Y.Z.; methodology, H.Z.; software, J.L.; validation, C.Y., S.C. and J.L.; formal analysis, Y.Z.; investigation, C.Y.; resources, H.Z.; data curation, Y.Z.; writing–original draft preparation, S.C.; writing–review and editing, J.L.; visualization, S.C.; supervision, Y.Z.; project administration, J.L.; funding acquisition, H.Z. All authors have read and agreed to the published version of the manuscript.

Funding: This work was supported by the National Natural Science Foundation of China under Grants Nos. 61873195 and 51707071.

Acknowledgments: The authors like to express sincere gratitude to School of Electrical Engineering and Automation, Wuhan University for making this research work for execution and technical implementation in real-time.

Conflicts of Interest: The authors declare no conflict of interest.

References

1. Uriarte, F.M.; Smith, C.; VanBroekhoven, S.; Hebner, R.E. Microgrid ramp rates and the inertial stability margin. *IEEE Trans. Power Syst.* **2015**, *30*, 3209–3216. [CrossRef]
2. Mendis, N.; Muttaqi, K.M.; Perera, S. Management of battery-supercapacitor hybrid energy storage and synchronous condenser for isolated operation of PMSG based variable-speed wind turbine generating systems. *IEEE Trans. Smart Grid* **2014**, *5*, 944–953. [CrossRef]
3. Lai, J.; Lu, X.; Yu, X.; Monti, A. Cluster-Oriented Distributed Cooperative Control for Multiple AC Microgrids. *IEEE Trans. Ind. Inform.* **2019**, *15*, 5906–5918. [CrossRef]
4. Lai, J.; Lu, X.; Yu, X.; Monti, A. Stochastic Distributed Secondary Control for AC Microgrids via Event-Triggered Communication. *IEEE Trans. Smart Grid* **2020**, to be published. [CrossRef]
5. Lai, J.; Lu, X.; Yu, X.; Monti, A.; Zhou, H. Distributed Voltage Regulation for Cyber-Physical Microgrids with Coupling Delays and Slow Switching Topologies. *IEEE Trans. Syst. Man Cybernet. Syst.* **2020**, *50*, 100–110. [CrossRef]
6. Lai, J.; Lu, X. Nonlinear Mean-Square Power Sharing Control for AC Microgrids under Distributed Event Detection. *IEEE Trans. Ind. Inform.* **2020**, to be published. [CrossRef]
7. Refaat, S.S.; Abu-Rub, H.; Sanfilippo, A.P.; Mohamed, A. Impact of grid-tied large-scale photovoltaic system on dynamic voltage stability of electric power grids. *IET Renew. Power Gener.* **2018**, *12*, 157–164. [CrossRef]
8. Byshkin, M.; Stivala, A.; Mira, A.; Robins, G.; Lomi, A. Fast Maximum Likelihood estimation via Equilibrium Expectation for Large Network Data. *Nature* **2013**, *31*, 1. [CrossRef]
9. Brockmann, D.; Helbing, D. The Hidden Geometry of Complex, Network-Driven Contagion Phenomena. *Science* **2013**, *342*, 1337–1342. [CrossRef]
10. Centola, D. The Spread of Behavior in an Online Social Network Experiment. *Science* **2010**, *329*, 1194–1197. [CrossRef]
11. Li, X. The role of degree-weighted couplings in the synchronous onset of Kuramoto oscillator networks. *Physica A Stat. Mech. Its Appl.* **2008**, *387*, 6624–6630. [CrossRef]
12. Wu, C.W. Synchronizability of networks of chaotic systems coupled via a graph with a prescribed degree sequence. *Phys. Lett. A* **2005**, *346*, 281–287. [CrossRef]
13. Sorrentino, F.; di Bernardo, M.; Garofalo, F. Synchronizability In addition, Synchronization Dynamics Of Weighed And Unweighed Scale Free Networks With Degree Mixing. *Int. J. Bifurc. Chaos* **2007**, *17*, 2419–2434. [CrossRef]
14. Leyva, I.; Sendinanadal, I.; Almendral, J.A. Explosive synchronization in weighted complex networks. *Phys. Rev. E Stat. Nonlinear Soft Matter Phys.* **2013**, *88*, 042808. [CrossRef]
15. Zhu, L.; Tian, L.; Zhu, C.; Shi, D. Effects of coupling-frequency correlations on synchronization of complete graphs. *Phys. Lett. A* **2013**, *377*, 2749–2753. [CrossRef]
16. Chavez, M.; Hwang, D.U.; Amann, A. Synchronization is enhanced in weighted complex networks. *Phys. Rev. Lett.* **2005**, *94*, 218701. [CrossRef]
17. Jalili, M.; Mazloomian, A. Weighted coupling for geographical networks: Application to reducing consensus time in sensor networks. *Phys. Lett. A* **2010**, *374*, 3920–3925. [CrossRef]
18. Qin, B.Z.; Lu, X.B. Adaptive approach to global synchronization of directed networks with fast switching topologies. *Phys. Lett. A* **2010**, *374*, 3942–3950. [CrossRef]
19. Dong, Z.; Tian, M.; Lu, Y.; Lai, J.; Tang, R.; Li, X. Impact of core-periphery structure on cascading failures in interdependent scale-free networks. *Phys. Lett. A* **2019**, *383*, 607–616. [CrossRef]
20. Tian, M.; Dong, Z.; Cui, M.; Wang, J.; Wang, X.; Zhao, L. Energy-supported cascading failure model on interdependent networks considering control nodes. *Physica A Stat. Mech. Its Appl.* **2019**, *522*, 195–204. [CrossRef]
21. Dorfler, F.; Chertkov, M.; Bullo, F. Synchronization in complex oscillator networks and smart grids. *Proc. Natl. Acad. Sci. USA* **2013**, *110*, 2005–2010. [CrossRef] [PubMed]
22. Bidram, A.; Davoudi, A.; Lewis, F.L.; Nasirian, V. *Cooperative Synchronization in Distributed Microgrid Control*; Springer: Cham, Switzerland, 2017.
23. Fazlyab, M.; Dorfler, F.; Preciado, V.M. Optimal Design for Synchronization of Kuramoto Oscillators in Tree Networks. *Physics* **2005**, *92*, 181–189.

24. Gu, Y.Q.; Shao, C.; Fu, X.C. Complete synchronization and stability of star-shaped complex networks. *Chaos Solitons Fractals* **2006**, *28*, 480–488. [CrossRef]
25. Dekker, A.H.; Taylor, R. Synchronisation Properties of Trees in the Kuramoto Model. *Siam J. Appl. Dynam. Syst.* **2012**, *12*, 97–100.
26. Zhai, M. Transmission characteristics of low-voltage distribution networks in china under the smart grids environment. *IEEE Trans. Power Deliv.* **2011**, *26*, 173–180. [CrossRef]
27. Dorfler, F.; Bullo, F. Synchronization in complex networks of phase oscillators: A survey. *Automatica* **2014**, *50*, 1539–1564. [CrossRef]
28. Simpson-Porco, J.; Dorfler, F.; Bullo, F. Synchronization and power sharing for droop-controlled inverters in islanded microgrids. *Automatica* **2013**, *49*, 2603–2611. [CrossRef]
29. Pagani, G.; Aiello, M. Towards decentralization: A topological investigation of the medium and low voltage grids. *IEEE Trans. Smart Grid* **2011**, *2*, 538–547. [CrossRef]
30. Wu, Y.; Lee, C.; Liu, L. Study of reconfiguration for the distribution system with distributed generators. *IEEE Trans. Power Deliv.* **2010**, *25*, 1678–1685. [CrossRef]
31. Li, C.; Chen, G. Network connection strengths: Another power-law? *arXiv* **2003**, arXiv:cond-mat/0311333.
32. Barrat, A.; Barth, M.; Pastorsatorras, R. The architecture of complex weighted networks. *Proc. Natl. Acad. Sci. USA* **2004**, *101*, 3747–3752. [CrossRef] [PubMed]
33. Dynamic IEEE Test Systems. 2019. Available online: https://www.kios.ucy.ac.cy/testsystems/index.php/dynamic-ieee-test-systems (accessed on 13 October 2019).
34. Power Systems Test Case Archive. 2013. Available online: http://www.maths.ed.ac.uk/optenergy/NetworkData/iceland/ (accessed on 13 October 2019).

MDPI
St. Alban-Anlage 66
4052 Basel
Switzerland
Tel. +41 61 683 77 34
Fax +41 61 302 89 18
www.mdpi.com

Energies Editorial Office
E-mail: energies@mdpi.com
www.mdpi.com/journal/energies